The Economics of Offshore Oil and Gas Supplies

The Economics of Offshore Oil and Gas Supplies

Frederik W. Mansvelt Beck
Karl M. Wiig

Arthur D. Little, Inc.

Lexington Books
D.C. Heath and Company
Lexington, Massachusetts
Toronto

Library of Congress Cataloging in Publication Data

Mansvelt Beck, Frederik W
    The economics of offshore oil and gas supplies

    Final report of a study undertaken by Arthur D. Little, Inc. for the
Bureau of Land Management.
    Includes index.
    1. Petroleum in submerged lands—United States. 2. Gas, Natural, in
submerged lands—United States. I. Wiig, Karl M., joint author. II. Little
(Arthur D.) Inc. III. United States. Bureau of Land Management. IV. Title.
TN872.A5M38   553'.28'0973   76-54558
ISBN 0-669-01306-4   Rev.

Published simultaneously in Canada.

Printed in the United States of America.

International Standard Book Number: 0-669-01306-4

Library of Congress Catalog Number: 76-54558

Contents

List of Figures

List of Tables

Preface

This study of future expected OCS production volumes and costs has been undertaken at a time when the majority of the OCS areas are largely unexplored and when little information is available on the resource base, the geology, or the production technologies which will be feasible. As a result, the methodology of this study has allowed for the projection of results under uncertainty with analogous information about conditions in overseas environments which may be similar to the new OCS areas.

Resource projections, based upon those available from the USGS at this time, are considered preliminary by the experts who have assembled them. New projections on critical information items such as oil and gas in place, field-size distributions, and well productivities are currently being prepared, but will not be available for some time to come.

The book is based on the final report of a study undertaken by Arthur D. Little, Inc. (ADL) for the U.S. Department of Interior, Bureau of Land Management, under Contract No. 08550-CTS-48 which was concluded during the summer of 1976. The Bureau of Land Management's contract monitor was Carl S. Pavetto whose support was partly responsible for the broad methodological approach to the study. Karl M. Wiig was ADL's Project Director while Frederik W. Mansvelt Beck acted as the Associate Study Director. A number of individuals contributed to this study: Susan G. Johnson, Richard L. Lacroix, Richard F. Messing, Philip J. O'Brien, and William S. Hawes. These individuals provided significant inputs to the analyses, and Malcolm E. Cloyd contributed to the data collection of this study. The resource estimates, upon which this study is based, were generated by the United States Geological Services. The assistance of Dr. Peter Rose and Betty Miller of USGS and their staff was of great help in understanding and utilizing these estimates. This work could not have been completed without the help of the support staff which Nancy J. Gray coordinated with great skills.

Frederik W. Mansvelt Beck
Karl M. Wiig
Cambridge, Massachusetts
June 1977

The Economics of Offshore Oil and Gas Supplies

1 Introduction

During the last few years, the United States has become increasingly concerned with its future energy sources in light of declining domestic production, increased demand for energy, environmental problems, the sharp increases in prices, and the decrease in the security of supply of imported oil and gas. The reason for this concern is best illustrated by the growing gap between projections of future demand and future supply deemed possible from domestic sources in existing areas as shown in Figure 1-1. As a result, both government and private industry have been focusing upon finding possible new sources of energy. The promising United States outer continental shelf (OCS) areas have been a major point of interest due to their high potential as a large source of oil and gas. To a large

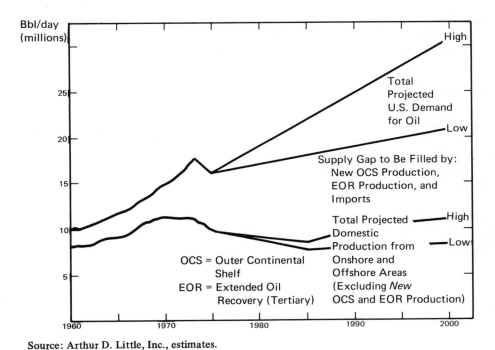

Source: Arthur D. Little, Inc., estimates.

Figure 1-1. Projections of Domestic Oil Supply and Demand.

1

extent, OCS areas around all United States coastlines consist of sedimentary rocks of the general types in which oil and gas are normally found. As a result, oil and gas production may be possible from all the 17 areas into which the United States OCS is divided, with some yet unexplored areas being very likely to contain large amounts of commercially producible oil and gas. There are significant questions, however, about which areas contain the best prospects, with regard to both the magnitude of attainable production streams and the landed costs of the available oil and gas. In addition, there are questions on the requirements of the capital which will be needed to find, develop, and produce these resources.

The finding, development, and production of oil and gas from the OCS in the United States is performed by the private sector; individual companies lease on individual blocks from the federal government the rights to perform these activities. The Bureau of Land Management (BLM) of the U.S. Department of the Interior manages this leasing process and offers for lease selected, promising tracts that are nominated by private industry which has performed some geological and seismic exploration of the area. Leases are awarded through competitive bidding where the winning bid prices often directly reflect the perceived potential of a block and may account for significant front-end investments of hundreds of millions of dollars for the rights to explore one block.

Since federal OCS leasing began 22 years ago, about 13 million acres have been leased altogether, with by far the greater part of this acreage in the Gulf of Mexico.

The federal government received $609.6 million in 1975 from royalties on oil and gas production on the outer continental shelf, according to the Interior Department's U.S. Geological Survey. OCS royalties ($597.2 million of which is from tracts in the Gulf of Mexico and $12.4 million from offshore California tracts) represent 69 percent of total royalties collected by the United States in 1975 for energy exploration and production on federal lands. Offshore also represents 1812 leases, covering 8.4 million of the 101.4 million acres leased by the federal government for oil and gas production. In 1975, more than 13,500 producing leases were estimated to yield 593 million barrels of crude oil and natural gas liquids and 4.5 million MCF of marketed gas valued at more than $6.4 billion. This yield represents 22.4 percent of the marketed gas and 16.2 percent of total crude and gas liquids produced in the United States during 1975.

OCS Economics and Costs

The process of exploration includes all the steps necessary to locate potential sources of petroleum and to establish their presence in commercial-size accumulation. On the OCS, this may involve, among other activities, the drilling of one or more exploratory wells for each geophysical prospect. Exploratory expenditures for drilling in 200-meter water depths and in moderate climatic conditions, such as found in the Gulf of Mexico, may amount to approximately $3 million for a 10-000-foot well. These are the purely technical costs incurred for exploratory well drilling. It should be recognized that these costs do not include other significant offshore exploratory expenditures such as lease bonuses, geological costs, and certain overhead expenditures that will normally be allocated to the exploratory effort.

As exploratory drilling progresses to greater water depths and more severe climates, drilling expenditures will necessarily increase. The primary factors contributing to these increased expenditures are the rig capital costs and the drilling and equipping time involved.

The market increase in costs as a function of water depth and climatic severity also applies to development and production expenditures. In water depths where sea floor producing units can be utilized, the cost of producing facilities is not expected to show a sensitivity to increasing water depths to the same extent as in the water depths range where platform-type installations can be employed. Of course, the distance from shore will contine to affect expenditures. In contrast with the exploration activity, which usually requires very few wells, the commercially successful offshore field requires a large number of development wells together with associated gathering, separating, storage, and transportation facilities, including safety and environmental protection facilities. For the moderately severe climate, such as the Gulf of Mexico, and for water depths not exceeding 150 meters, the total cost for a development and production system can be estimated at approximately $125 to $150 million for a 100-million-barrel oil field. In other words, this is a production system that will operate under the same conditions as an exploratory well that can be drilled for $3 million. If climatic conditions become severe, such as can be found, for instance, in the Gulf of Alaska and Lower Cook Inlet, the costs for the same production system may be 3 times as much, i.e., in the range of $375 to $450 million. Increased water depth, especially in areas with severe climatic conditions, may increase the costs as much as threefold, i.e., if the water depth is increased from 150 to 200 meters.

For 75 percent ice-laden areas, drilling in deep open waters may be possible only during 3 to 4 months of the year. In areas with severe climatic conditions, platforms are not assumed economically feasible beyond 200-meter water depth; therefore, floating/drilling, together with sea floor producing systems, is required. The cost of those producing systems will be substantially above the cost of systems in what is now considered severe climatic conditions and deep waters.

Objective of Analysis

The analysis presented on the following pages was undertaken to support the Bureau of Land Management in its efforts to establish the potential of OCS oil and gas. The objectives of this study are to project the future oil and gas costs and production streams for all 17 areas resulting from available resource estimates, as laid down in the Lease Planning Schedule of June 1975. Also, the capital requirements for support of exploration, development, and production are analyzed.

Methodological Approach

The methodology which was developed concentrated on allowing for differences between the various areas in the quality and quantity of the resources, the access to those resources and exploration, and development and production costs, all of which are of prime importance to policy makers who control the general access to those areas.

The main direction of the methodological approach is based upon the notion that production volumes, unit costs (dollars per/barrel, dollars per thousand cubic feet), development capital, and time requirements are very sensitive to the size of individual fields encountered in the area under analysis. As a consequence, the analysis performed for this study projects the *size* and rate of fields found and developed each year as a function of the areas leased and the resulting exploration activities. For such an analysis to be valid, the technical costs for each required activity are assembled based upon the technology which is forecasted to be employed for a particular field size in a particular OCS area in a given year to produce annual cash flow streams associated with the ultimate production of a field. Since an OCS area may contain a number of fields, some of which are being developed and produced simultaneously, all costs and production streams are aggregated to allow projection of average values of unit costs and production volumes from an area. From the analysis, "minimum economic field sizes" are projected for the individual areas. When the OCS costs are projected throughout this study, any economic rent—in terms of lease costs—was excluded to yield the minimum possible costs of oil and gas from the OCS.

There is considerable uncertainty associated with the total resources available from OCS areas, as well as with how these resources may be distributed over different-sized individual fields. In order to project the total expected costs and production streams, a large number of equally likely scenarios were simulated of distributions of resource estimates, structure sizes, and the degree to which the structures are filled with hydrocarbons.

A data base was assembled from private industry sources on possible technologies and their costs, from government agencies and private organizations on supply-and-demand projections and past exploration experiences, and from U.S. Geological Survey on OCS resource projections.

An economic assessment of the value of the OCS resource must take into account the time at which expenditures are made and production becomes available. Hence, the assessment must include projection of the industrial dynamics of the exploration and development process and must allow for the fact that companies, which are operating in a certain area, are competing for a limited number of workers, materials, and equipment.

There is some degree of uncertainty associated with most of the cost elements for exploration and development activities, all of which makes the overall prospecting and policy environment risky and difficult to manage.

In order to provide a suitable means for projecting impacts of alternate scenarios about this uncertain environment, a methodological framework was developed to allow for:

1. Inclusion of uncertainties surrounding the resource base estimates, assumed field size distributions, unit equipment and operating costs, exploration and development durations, and inflation rates
2. Presentation of the results of different assumptions about the availability of workers, materials, and equipment to sustain the exploration and development effort

This adopted approach permits the assembling of aggregate cost estimates from the individual estimated costs. The individual estimates are made for physical equipment units, such as platforms, and production equipment and unit activities, such as the daily drilling contractor costs. Estimates, made for disaggregated cost elements, ensure that cost differences, resulting from varying conditions, such as water depth in the case of platform costs, can be specified separately for each of these physical units and unit activities. Further, disaggregation recognizes the variability in inflationary tendencies among cost elements. The result is an accurate estimate of aggregated costs. This approach further provides a basis for analyzing the sensitivity in production forecasts as a result of making different assumptions about potential recoverable reserves and field size distribution. In addition, the ranges and likelihoods of occurrence of different important cost measures, as functions of specific production scenarios, have been analyzed with indications of how these costs will be different for different external conditions.

2

Methodology for Projecting Oil and
Natural Gas Production and Costs
from the United States Outer
Continental Shelf

Overview

Only a relatively small part of the outer continental shelf (OCS) of the United
States has been explored. There is a high degree of uncertainty about the levels
of oil and gas resources which might be present in the different remaining unexplored OCS areas, the distribution of these resources over different-sized fields,
and the production characteristics of individual fields.

The basic approach of the selected methodology as discussed in this chapter
is to project the total costs associated with *each individual field* on an annual
basis associated with the resulting production streams. This approach has been
chosen since the experienced cost per unit of production is strongly coupled
with the size, technology, and geographical context of *individual fields* and
since it is expected that a wide range of oil and gas field sizes will be discovered
and produced under a wide range of possible circumstances in each OCS area.

An overview of the analysis methodology and its information flow is presented in Figure 2-1. The analysis is built around a set of computer-based
models which:

1. Simulate over a planning horizon beyond 1990 the dynamics of the exploration, development, and production process of an OCS area, in general, and
   each explored field, in particular, subject to equipment availability con-
   straints
2. Build up accrued costs according to type of expenditures for each field
3. Project annual production of oil and gas of each field based upon its size
   and development program
4. Account for uncertainties by use of Monte Carlo simulation to simulate
   each OCS area a large number of times, each time with a different, equally
   likely scenario which is sampled from probability distributions by which the
   uncertain variables are expressed
5. Project the United States total energy supply-and-demand balance by state,
   with and without OCS oil and gas production for estimation of its impact

Parameters of the Problem

The major objective of this study is to project future oil and gas production and
their associated costs on selected tracts of the United States outer continental

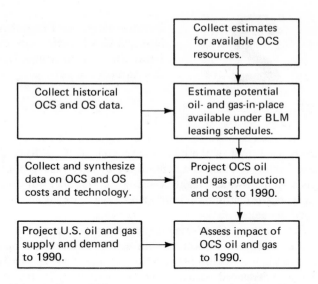

Abbreviations: OCS = outer continental shelf; OS = offshore.

Figure 2-1. Overview of Methodology.

shelf. For these purposes, the following four sets of parameters have to be estimated:

1. An approximation of the *total resource base* of oil and gas in the general area of the specified tract and an estimate of the portion of this total resource base that may underlay the tract itself.
2. A definition of the *quality* of the resource base in terms of both its expected field size distribution and its concurrent parameters of producibility of the trapped hydrocarbons, i.e., depths of producing horizons and well productivities.
3. A description of the physical *environment* of the tract as to prevailing weather conditions in the area, water depth, seasonal weather patterns, and other parameters which are necessary to assess the type of technology required for exploration and field development.
4. An assessment of both the *available technology* and its cost for exploration, development, and production of the specified tract, given its quality and its physical environment.

Although a correct analysis of potential production requires the determination of all the parameters indicated above, most of the values are not known with certainty. There are different levels of uncertainty associated with different parameters; therefore, a methodology has to be adopted to quantify these uncertainties and, subsequently, to aggregate the uncertain variables and parameters into estimates of production and production costs. As a consequence of the probabilistic (uncertain) nature of the input data, it can be expected that the projected production levels and costs will be equally uncertain, and hence that they must be defined in a probabilistic sense.

Some parameters can be determined with a higher degree of confidence than others. The uncertainties present in the estimates of the quality of the resource base in terms of field size distributions and production characteristics of the fields themselves are much larger than the uncertainties inherent in the estimates of exploration and development costs for specific tracts. To allow for this difference in levels of uncertainty, the following two-step approach was used in forecasting potential production levels and their associated costs in the 17 different OCS areas considered:

Step 1. Using an aggregation procedure for probabilistically defined information, determine expected production profiles.

Step 2. In a deterministic sense, calculate the production cost of the specific tract under consideration.

Step 1 utilizes geological and oil industry information, including the uncertainty surrounding major variables, such as:

the size of the area which can be expected to be leased by companies, i.e., a tentative definition of the area to be leased

the size of the total resource base

the expected field size distributions

the exploration and development programs which companies can be expected to undertake

Uncertainties inherent in estimates of both the size of a resource base and field size distributions are so large that they should be properly allowed for when used to derive projections of possible production levels. This was achieved by the use of the methodology of stochastic (or Monte Carlo)[1] simulation, which allows for the aggregation of probability distributions of complex processes. The application of this methodology results in a probabilistic, rather than a deterministic, estimate of resulting oil and gas production from opening OCS areas to exploration and development. Coupled with the results of step 2, an estimate can then be derived of how much capital may be required over future years to sustain these exploration and development efforts.

In order to perform step 2, exploration, development, and production costs have been developed for a certain field size, as a function of:

the required technology

environmental parameters (water depth, distance to shore, etc.)

the capital cost of the industry

the fiscal regime of oil and gas production in the United States

These cost functions take the form of the example in Figure 2-2. It will be readily apparent that, under a given set of assumed parameter values, there is a minimum field size below which development becomes uneconomic under prevailing market prices for oil and gas. The determination of this "minimum economic field size" under current economic conditions and for the various areas of the OCS is an important result of this study. A more detailed description of the issue of minimum economic field size can be found in the section Minimum Required Price.

Prices for oil and gas can change dramatically in the lapse of time between discovery of a new field and first production. This implies that companies will have to make their decision to go ahead and develop a newly discovered field based on their projections of price and cost levels expected possibly as much as 7 years hence. If they expect future prices in real terms to be higher or costs to be lower than at the time of discovery, those fields that are considered to be submarginal under existing price/cost conditions will still be developed. On the other hand, if companies expect future prices to be lower or costs to be higher than at the time of discovery, then fields might not be developed which would be economical to produce under present cost/price conditions. For the purposes of this study, projections have been made of probable future production levels under different cost/price scenarios to indicate the increase in production which can reasonably be expected at higher than current price levels.

Resource Estimations

In general, only part of all the oil and gas fields present in a particular area is made accessible for exploration drilling through a lease sale; private companies bid for rights to explore and produce on specific tracts. The bids may be based on good information obtained through seismic investigations about the presence of structures, but at the time of the bid no substantive information is available about the presence of oil or gas in the structures. The best a potential bidder can do is to assume the presence of oil and gas based on analogies with resource bases that have been developed already. Consequently, companies, in their

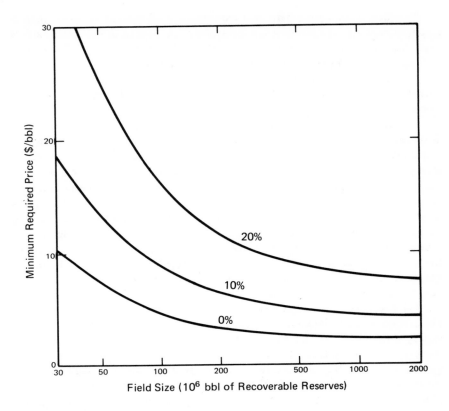

Source: Arthur D. Little, Inc., estimates.

Minimum Required Price as a Function of Field Size—Oil
Average Well Productivity 2500 bbl/day
Water Depth: 400 feet
Distance to Shore: 25 mi
Number of Years Delay: 5

Figure 2-2. Required Price Schedules, Gulf of Alaska (1975 Dollars).

bidding procedures, will first concentrate on so-called structural traps, the presence of which they know through their seismic work. Second, companies will concentrate on the larger structures, since these hold promise for the largest fields. In this analysis the industry dynamics of exploration, development, and production activities in offshore areas are simulated with allowance made for this particular aspect of offshore exploration in conjunction with the uncertainty of the geology of the resource base.

In order to simulate a particular exploration environment of a certain geographical area, e.g., the North Atlantic, estimates are needed of the total resource base which is expected to be present in the area and of the expected distribution of field sizes in that particular resource base.

One complete iteration in the simulation process, for a certain geographic area is carried out as follows (see also Figure 2-3) in order to create our resource scenario.

(1) Sample the distribution of expected total size of the resource base. This distribution reflects the uncertainties about the amount of oil and gas which may be present in an area. A typical distribution may take the form of the example in Figure 2-4 for oil. A similar distribution is available for gas. The result of this single sampling is the determination of the total amount of oil (and gas) present in the resource base for *this particular scenario*. The single sample is drawn at random but in such a fashion that the chance for a particular value to be drawn follows from the probability distribution, as shown in Figure 2-4. Conversely, this implies that if many samples are drawn, in subsequent iterations the total of all those samples will constitute a distribution such as the one in Figure 2-4, the probability distribution of the likely size of the total resource base for the area being analyzed.

(2) The next step, in the process of building up one scenario of a complete exploration and development program for a certain area, is to allocate the previously determined oil and gas resource base over structural traps of different sizes, which then contain oil, gas, or nothing. This is achieved by sampling the general structural trap size distribution as derived for this area, using estimates for the average success ratios in terms of the number of dry versus successful exploratory wells and for the number of wells required to explore a structural trap of given size and complexity. The conditional probability of the particular trap being dry is then established. Through use of the Monte Carlo technique, it is decided whether the trap is dry. If it is not dry, then the amount of oil or gas present in the trap is obtained through sampling of a *fill factor* distribution. Structures, if not dry, will be filled with oil or gas depending on which of the two remaining resource bases is the largest. The fill factor determines the average amount of recoverable oil and gas per unit area present in the trap; i.e., it is a proxy for the richness and recoverability of the oil or gas reservoir contained in the trap.

(3) At this point the total number of structural traps present in the area have been determined, some of which hold the entire resource base in terms of recoverable oil and gas while others are dry. The actually leased area, though, usually contains only a fraction of the total resource base present in the general area. Hence, as a final step, the leased area is filled with structures, starting with the largest; some will contain oil, or gas and others will be dry. The result is the establishment, through simulation, of a complete exploration and development environment scenario in the area under study, predicated upon the amount of resources assumed to be present in that area.

(4) This environment will be subjected to an exploration and, eventually, a development program. Total yearly production and the associated production

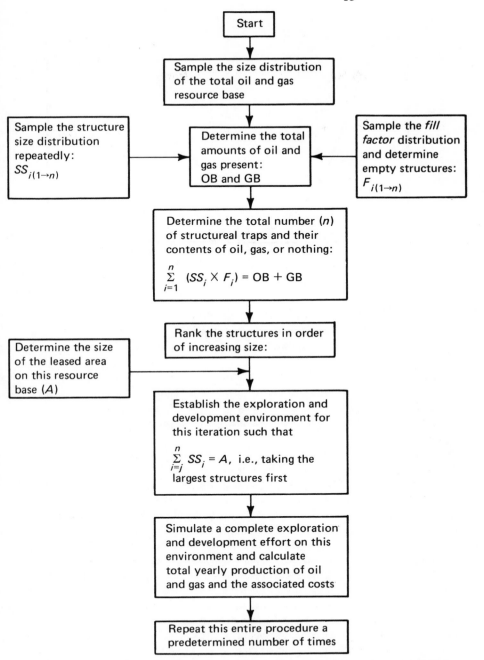

Figure 2–3. Simplified Flow Diagram of the Procedure to Simulate an Exploration and Development Environment in a Certain Geographic Area (One Iteration).

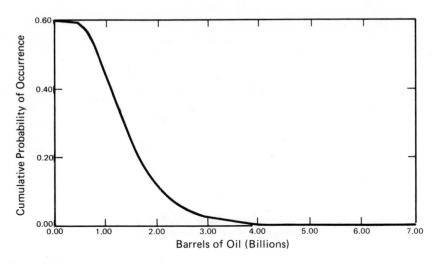

Source: U.S. Department of the Interior.

Figure 2–4. Probability Distribution of Occurrence of Undiscovered Recoverable Oil Resources in the North Atlantic Region of the Outer Continental Shelf (water depth 0-200 m).

costs will be calculated under various assumptions about the market prices of oil and gas. The latter assumptions are necessary since total production of a lease is a function of the price obtained for the oil and gas. High prices will render high cost production of small fields economically viable, production that would not be obtained at lower market prices.

This particular iteration is finished with the calculation of production levels and costs. The parameters and results of the calculations are stored, and the entire simulation procedure is repeated a large number of times to project the production streams and costs under equally probable scenarios. The results of all iterations are distributed according to frequencies of occurrence, e.g., as shown in Figure 2-5.

Figure 2-5 can be interpreted as follows. After a total of 100 iterations (100 scenarios), the calculation procedure obtained cumulative production between 1 and 2 billion barrels in five iterations, i.e., in 5 percent of the cases. It obtained cumulative production of between 4 and 5 billion barrels in 28 of the 100 iterations, i.e., in 28 percent of the cases. These results can also be expressed as, for instance, a chance of 28 percent to obtain cumulative production between 4 and 5 billion barrels. Another way to express the results is in a cumulative sense; e.g., there is a 95 percent chance of obtaining cumulative production of 2 billion barrels or less; there is an 86 percent chance [100 – (5 + 9)] of obtaining 3 billion barrels or less, etc. This can be shown graphically as a cumulative distribution (Figure 2-6). The cumulative production of 2 and 3 billion

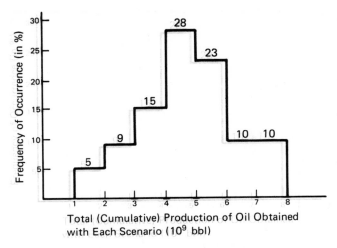

Figure 2-5. Frequency Distribution of Cumulative Production from 100 Scenarios.

barrels in this example are also said to be the production at the 95 and 86 percent "confidence levels," respectively.

In the analysis itself, annual production and capital expenditure levels, as might result from an accelerated lease sale schedule through 1978, are projected for the different OCS areas at confidence levels of 5, 25, 50, 75, and 95 percent, respectively.

Simulation of Exploration and Development Effort

Once the exploration environment of the leased area has been defined in terms of a list of various-sized structural traps, each one of which either is filled with oil or gas or is empty, ADL's basin development model is used to simulate the subsequent exploration and development effort. For this iteration, this simulation results in a determination of the expected value of total production and of total production costs at various price levels for oil in the United States market. Price level scenarios are a necessary condition for production and cost calculations since ultimate recovery depends upon the price obtained for the marginal barrel. As stated earlier, high prices will justify the recovery of high-cost oil and thus will effectively increase total production of a given resource base; the reverse reasoning holds true for low prices.

A complete exploration and, if successful, a subsequent development effort are then simulated in the following chronological steps (see also Figure 2-7):

(1) Take the largest structure off the list of structures underlying the area. Determine its distance to shore, the water depth at its location, and the depth of the target formation.

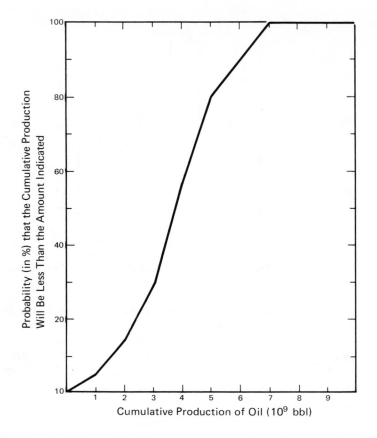

Figure 2-6. Cumulative Distribution of Cumulative Production Which May Be Expected from the Area

(2) Drill exploratory wells. The results of this drilling should help determine whether the structure contains any commercially producible oil or gas or whether it is dry. The number of exploration wells to be drilled depend on:

the size of the structure

the method of development of the structure, that is, whether it is developed tract by tract or whether companies owning the tract pool their exploration efforts (pooled efforts usually result in a smaller number of exploratory wells, since information on them is exchanged)

the ratio of the area containing reserves to the total area of the structure

Drilling of these exploration wells is constrained by the availability of exploration rigs in the total leased area in any year. Given estimates of how long it will

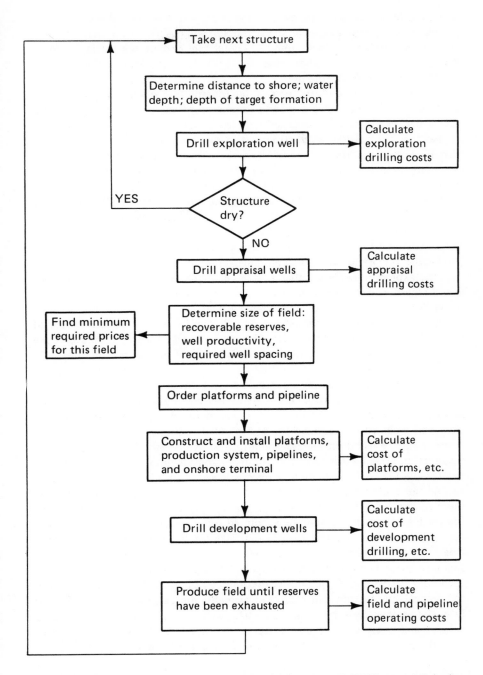

Figure 2-7. Simulation of the Exploration and Development Effort and Calculation of Associated Costs and Production for a Leased Area.

take to drill an exploratory well in the area and how long the exploration drilling season is expected to be, the simulation will calculate how many years of exploration drilling will be required to completely explore the particular structure under analysis. The costs associated with the exploration drilling are calculated and stored for future use.

(3) If the structure is dry, the next largest structure will be taken off the list of structures for the leased area, and simulation of exploration drilling on this new structure will be done in exactly the same manner as described before. This implies that exploration on all structures in the leased area will start in the first year if enough rigs are present in the area to drill each of these structures. If not enough rigs are present in the area, then drilling will have to wait until drilling on other structures has been finished. In that case, the largest structures in the leased area are drilled first because the simulation selects them first, reflecting the fact that the industry will want to know whether the largest structures in a leased area contain any oil or gas before spending their exploration dollars on the smaller structures which have a lower chance of containing economically producible oil or gas. If a predetermined number of successive structures are dry, the exploration effort in that area will be halted because companies will not spend any more exploration dollars if the chances of finding oil or gas in remaining undrilled structures become increasingly small.

(4) Subsequently, a number of appraisal wells are drilled, the timing of which depends upon rig availability. Exploratory and appraisal drilling are done with the same type of rig. Appraisal drilling is performed on structures with proved reserves, aiming at delineating the field contours. Again, the costs associated with appraisal drilling are calculated and stored for future use.

(5) If the explored structure contains any oil or gas, the production characteristics in terms of the average well productivity, depth of producing formation, and well spacing required for development wells are specified. This specification can be based on average conditions of similar structures or on specific information available for the structure under consideration. The information on recoverable reserves, production characteristics, and location of the field is used to find the minimum required wellhead price for oil that will justify the development of this field. This is achieved by means of previously determined functions (determined through the minimum required price analysis described in the next section) that relate required price to field size for fields in certain geographical areas and of a certain "quality." Figure 2–8 shows an example of one of these functions with identification of the parameters governing the functions. Since each field has its own minimum wellhead price below which development is not economical, production and capital investments are categorized for each field into classes identified by their minimum wellhead price. Hence, production volume and costs are put into categories with wellhead prices, for instance, of $10 per barrel or lower, $15 per barrel or lower, $20 per barrel or lower, etc.

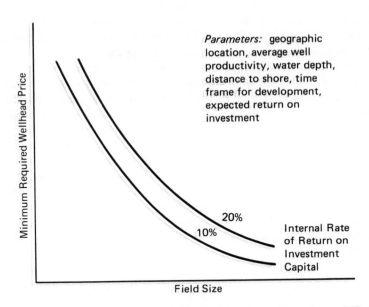

Figure 2-8. Minimum Required Wellhead Price as a Function of Field Size (Results from Minimum Required Price Analysis).

(6) After the drilling of appraisal wells is simulated to delineate the field contours, the procedure projects the number of development wells required to produce all the recoverable reserves of the field and the number of development platforms needed to accommodate processing facilities for the expected production of the field. The capacity of the transportation system to bring the production to shore and the size of the shore terminal are then determined.

In the case of oil, the transportation system can consist of a pipeline or a tanker loading facility. In the case of a pipeline, depending on the field size, the pipeline can be dedicated for this field alone if the field is very large, or for linking up the field production with a larger pipeline to shore accommodating production from different smaller-sized fields. The simulation allows for the fact that pipeline and platform construction and installation usually take more than a year. The number of platforms that can be constructed in any given year is limited by the number and capacity of construction sites. The number of years required for appraisal and development of a field and the number of years between first discovery and first production are thus correctly simulated as being dependent on not only the location of the field, the size of the field, and the production characteristics, but also the assumed or expected availability of drilling rigs and platform construction sites.

The costs associated with pipeline construction, with platform construction and installation, and with the construction of the gathering system linking up the different platforms in a field with the pipeline to shore are calculated separately by procedure as these activities are simulated over time.

It should be noted that production costs, as calculated at this point for a given field, are similar to the costs that were calculated to find the minimum economic wellhead price. However, the latter calculation was carried further into a complete cash flow analysis, taking into consideration the applicable tax burdens.

(7) Once platforms are installed, development drilling can start. Again, it will take time until all development wells have been drilled from a platform. After development wells have been drilled for a given platform and the transportation system has been constructed, production from the field can begin. The total production profile for the field consists of the production profiles of the individual wells as they are brought into production after having been drilled. Operating costs associated with field production are also calculated and stored separately.

Once the simulation of exploration, appraisal, development, and production efforts for the given field has been finished and the associated costs and production profiles have been calculated and saved, the next structure is selected from the list of structures expected to be present in the area under the present scenario; then the entire simulation is repeated. This cycle will continue until the list of structures included in the leased area is completely exhausted.

In case several structures contain oil or gas, the simulation of the exploration, development, and production activities in the area will result in production and capital expenditure profiles over time for different minimum price categories. In other words, the procedure will have calculated how much annual oil and gas production and associated costs can be expected at different levels of future oil and gas prices if, indeed, the size of the resource base and its allocation over different-sized structures are assumed in this particular iteration.

After a large number of iterations, the procedure will have developed a like number of production and capital expenditure profiles for each of the assumed price categories where the production and capital expenditure profiles will range from zero (if there is a chance that the area does not contain any oil or gas) to the highest production volume which might be possible if, indeed, the largest estimated amount of oil or gas is present in the area.

Minimum Required Price

Some of the fields that might be found in the different outer continental shelf areas considered in this analysis, if developed, will be profitable under the present cost/price conditions. Other fields, especially the smaller ones, might not be profitable with present-day prices and costs. In addition, prices and costs may change relative to one another in the future.

Companies that decide to go ahead and develop a particular field in a particular area will base their decision on what they expect price/cost relationships to be over the life of the field. The purpose of this study was *not* to make a forecast of what price/cost relationships for offshore field exploration and development can be expected to be. The purpose was to show which production levels could be expected if future prices assume certain prespecified values relative to the cost of exploration, development, and production. For this purpose, the "minimum required price" concept has been used. The *minimum required price* is that constant price over the life of the field at which field production will pay for the development and operating costs of the field with an allowance for royalty and tax payments and for the company's capital costs.

As shown in Figure 2-9, the minimum required price is the break-even price for which the present value of total yearly revenue (production times price) is equal to the present value of all outlays, i.e., exploration drilling costs, investment costs in field development, and field operating costs, once royalties and tax payments have been taken into account. Thus, the minimum required price is calculated as if companies had perfect information about the size and quality of the field which they are going to find, the exploration drilling costs, and the development and production costs to be incurred to bring the field into production.

Companies make this type of calculation previous to a lease sale, when they have to decide to bid or not to bid on a particular block. In other words, if they find that the minimum required price for the expected field size is higher than expected future price levels, most probably they will not bid on that particular block. If they find the minimum required price for the expected field to be lower than estimated future price levels, they may bid on the block. *Their maximum bid can be expected to be the difference between the minimum required price as calculated and the perceived future oil price.* In theory, society would thus reap the maximum economic rent. On the margin, where the minimum required price as calculated for the expected field is equal to the estimated future price levels, they can be expected to bid with a zero cash bonus. The minimum required price, therefore, allows for showing what the smallest field size is in a particular area which companies would be willing to look for at that price level.

Exploration costs were considered only nominally in the calculation of the minimum required price. Costs included in the minimum required price calculation were only costs required for field exploration, development, and subsequent production as estimated to be necessary when making the bid/no-bid decision prior to the lease sale. Expenditures which have already been made, i.e., the seismic and geophysical survey costs and the exploration drilling costs for dry

Without Inflation/Deflation

Step 1: $PV[(PRICE \times PROD) \times (1 - TAX) \times (1 - RYLTY)] =$
$$PV(EXP_{tax}) + PV(DEV_{tax}) + PV(OC_{tax})$$

Step 2: $$PRICE = \frac{PV(EXP_{tax}) + PV(DEV_{tax}) + PV(OC_{tax})}{PV[PROD \times (1 - TAX) \times (1 - RYLTY)]}$$

With Inflation/Deflation

Step 1: $PV[PRICE \times INFL \times PROD) \times (1 - TAX) \times (1 - RYLTY)] =$
$$PV(EXP_{tax}) + PV(DEV_{tax}) + PV(OC_{tax})$$

Step 2: $$PRICE = \frac{PV(EXP_{tax}) + PV(DEV_{tax}) + PV(OC_{tax})}{PV[PROD \times INFL \times (1 - TAX) \times (1 - RYLTY)]}$$

PV	= Present value operator
PRICE	= Minimum required price
PROD	= Annual production
TAX	= Tax rate
RYLTY	= Royalty rate
EXP_{tax}	= After-tax exploration expenditures (having allowed for deductibles)
DEV_{tax}	= After-tax investment and expenditures in field development
OC_{tax}	= After-tax field production costs
INFL	= Annual rate of change in PRICE relative to exploration, development, and production costs

Figure 2-9. Calculating the Minimum Required Price (= Price).

blocks, were not included. The latter expenditures can be regarded as a necessary cost of being in the oil business. The total return on capital has to be sufficiently high to repay these costs. Hence, it is the analyst's assumption on required returns which results in a more or less correct calculation of minimum required prices. The analysis is performed with various assumed rates of return, allowing some insight into the sensitivity of this variable.

Figure 2-2 shows a minimum required price schedule for the Gulf of Alaska. This schedule allows for an estimation of what the minimum economic field size will be if companies expect future prices to assume certain levels relative to the estimated field development and production costs. Using these price schedules, which were developed for the minimum economic field size analysis, the production for different-sized fields plus the capital expenditures required for exploration and development of these fields under different price scenarios was categorized. This categorization was done on two levels (see Table 2-1):

1. The projections of future potential production levels and associated capital expenditures were grouped into classes of increasingly higher prices for oil and gas.
2. Within each class, the probability of reaching certain levels is indicated by noting confidence levels between 5 and 95 percent.

Projections of Future Oil and Gas Production from Onshore Areas and Existing Offshore Areas

In order to assess the relative importance of expected production from new OCS areas, a forecast was made of future potential production from onshore areas and from existing offshore areas at the state level. For this purpose, mean values for estimated undiscovered recoverable resources for 75 petroleum provinces, as obtained from the USGS, were assigned to the individual states. A high and a low projection were made of total production by projecting separately:

Table 2-1
Expected Oil Production of the Eastern Part of the Gulf of Alaska
(10^6 Bbl/Year)

		1976	1980	1985	1990
Assumed Price: $4.50/bbl					
Confidence level:	5%	0	0	0	0
	25%	0	0	0	0
	50%	0	0	0	0
	75%	0	0	0	0
	95%	0	0	0	0
Assumed Price: $7.50/bbl					
Confidence level:	5%	0	0	359.28	258.02
	25%	0	0	80.88	60.38
	50%	0	0	0	0
	75%	0	0	0	0
	95%	0	0	0	0
Assumed Price: $12/bbl					
Confidence level:	5%	0	0	391.72	284.56
	25%	0	0	102.80	71.56
	50%	0	0	35.34	26.34
	75%	0	0	0	0
	95%	0	0	0	0

1. production from existing reserves
2. production from reserves added through revisions and extensions to reserves existing in 1974
3. production from newly discovered reserves
4. production from reserves resulting from extensions and revisions to newly discovered reserves

An "optimistic" and a "pessimistic" production forecast were made in order to establish a range within which the actual future production levels could reasonably be expected to fall.

The *optimistic production forecast* was obtained by assuming that economic incentives would result in an increase in discovery rates relative to 1974 levels. Under that scenario half (50 percent) of the undiscovered resources were assumed to be discovered within the next 25 years, and all the undiscovered resources were assumed to be discovered in the next 50 years.

The *pessimistic production forecast* resulted from assuming that a lack of economic incentives would result in relatively low future annual discovery rates, remaining at approximately the same level as realized in 1974.

Note

1. For an explanation of the technique of Monte Carlo simulation, the reader is referred to J. M. Hammersley and D. C. Handscomb, *Monte Carlo Methods*, Methuen & Co. Ltd., London (or Wiley, New York), 1964.

3

Data Base

Geographic Information

Outer Continental Shelf Geographical Divisions

The Bureau of Land Management (BLM) of the U.S. Department of the Interior has divided the outer continental shelf (OCS) of the United States into 17 different geographic areas. These are presented in Table 3-1.

Throughout this analysis, areas 5 and 6 (Central Gulf of Mexico and South Texas) have been consolidated since their oil and gas resources were estimated as one area by the USGS in the source material used for this study. The OCS areas have been consolidated into seven major areas for summary of production projections, as shown in Table 3-2.

OCS Geography

Based upon information published in the OCS Environmental Impact Statement (U.S. Department of Interior, "Final Environmental Impact Statement Proposed Increase in Oil and Gas Leasing on the Outer Continental Shelf"), the locations of the most significant structures are known for the 17 OCS areas. The estimates of the water depth and distance to shore are given in Table 3-3 for the seven consolidated OCS areas.

Resource Definition

Substantial work has been performed by the U.S. Department of the Interior on the estimation of undiscovered recoverable oil and gas resources in the United States. The results of that work have been presented in the Geological Survey Circular No. 725[1] which was prepared for the Federal Energy Administration in 1975. Estimates contained in Geological Survey Circular No. 725 are based on the expectations of geologists and geophysicists about the amounts of oil and gas that can be expected to be present in each different OCS area. These are made in a probabilistic sense, showing the chances that exist for different amounts of recoverable reserves of oil and gas to be present in the area.

25

Table 3-1
OCS Geographic Areas

OCS Area Number	Designation
1	North Atlantic
2	Mid-Atlantic
3	South Atlantic
4	MAFLA (Eastern Gulf of Mexico)
5	Central Gulf of Mexico
6	South Texas
7	Southern California
8	Santa Barbara Channel
9	Northern California
10	Washington-Oregon
11	Lower Cook Inlet
12	Gulf of Alaska
13	Southern Aleutian Arc
14	Bristol Bay Basin
15	Bering Sea
16	Chukchi Sea
17	Beaufort Sea

Table 3-2
Consolidation of OCS Areas for Production Summaries

Consolidated Area	OCS Area
Atlantic Coast	Areas 1, 2, and 3
Gulf of Mexico	Areas 4, 5, and 6
Pacific Coast	Areas 7, 8, 9, and 10
Gulf of Alaska	Areas 11, 12, and 13
Bering Sea	Areas 14, 15, and 16
Beaufort Sea	Area 17

Table 3-3
Estimates of Expected Water Depths and Distances to Shore for Consolidated OCS Areas

Consolidated OCS Area	Water Depth (ft)	Distance to Shore (mi)
1. Atlantic Coast	400	75
2. Gulf of Mexico	400	75
3. Pacific	600	15
4. Gulf of Alaska	400	25
5. Lower Cook Inlet and Bristol Bay	200	15
6. Bering and Chukchi Sea	200	75
7. Beaufort Sea	300	15

Although the USGS specialists did assume that the resources would be present in structural traps and that they would be present in fields large enough to make recovery technologically and economically feasible, they did not include an assessment of the number and size of fields in these estimates, in spite of the fact that most of the areas considered for future exploration have already been explored through seismic surveys. The information contained in these surveys, however, is not available for public review because it is the basis for evaluation and bid decisions of the very companies that performed the surveys. In most of the areas, there is some indication of larger structures which are believed to be present. For instance, in the Gulf of Alaska a very large structure is reported to be present in the Icy Bay area, and in the mid-Atlantic a structure of 72 square miles is believed to be present.

To a large degree the present study relies upon the source material for Circular No. 725 for its information on the likely probability distributions for the oil and gas resources of the OCS. These estimates were made through a review of geological and geophysical information on more than 100 different petroleum provinces in the United States and by applying a subjective methodology for estimation of the resources of each potential petroleum province. These resource appraisals were based upon group assessments by geologists and geophysicists and upon the application of subjective probability estimates of the various parameters. Monte Carlo simulation was used to provide aggregate estimates of the sizes of the resource bases underlying the 17 OCS areas as defined by BLM. Appendix A contains the 17 resource distributions of oil and of gas that pertain to the 17 outer continental shelf areas as defined by the BLM and as used in this study.

A summary of the oil and gas resource estimates used is shown in Table 3-4 for the 17 OCS areas in terms of their mean and subjective low and high estimates. The high estimate is specified at the 5 percent level of confidence; i.e., there is only a 5 percent (1 in 20) likelihood that the actual resources, when found, will exceed the high estimate. The low estimate is specified at the 95 percent level of confidence; i.e., there is a 95 percent (19 in 20) likelihood that the actual resources, when found, will exceed the low estimate.

Field Size Distribution

Since no explicit information is available on field size distributions in the new OCS areas, and since it is questionable if new areas will have distributions which may be similar to analog areas or areas with similar geology, *it was assumed that field size distributions in new areas could be approximated by the empirical United States average field size distribution for oil and gas fields.* Figure 3-1 shows the distribution of field sizes of the previously discovered fields in the

Table 3–4

Estimates of Undiscovered Recoverable Oil and Gas Resources in United States Offshore Areas

Water Depths of 0 to 200 Meters (Includes State and Federal Lands)	Crude Oil (10^9 bbl)			Natural Gas (10^{12} CF)		
	95 Percent Probability	5 Percent Probability	Statistical Mean	95 Percent Probability	5 Percent Probability	Statistical Mean
1. North Atlantic	0	2.5	0.9	0	13.1	4.4
2. Mid-Atlantic	0	4.6	1.8	0	14.2	5.3
3. South Atlantic	0	1.3	0.3	0	2.5	0.7
4. MAFLA (Eastern Gulf of Mexico)	0	2.7	1.0	0	2.8	1.0
5. Central Gulf of Mexico	2.0	6.4	3.8	17.5	93.0	44.5
6. South Texas	0.4	2.1	1.1	0.4	2.1	1.1
7. Southern California	0.6	3.0	1.5	0.7	3.3	1.7
8. Santa Barbara Channel	0	0.8	0.3	0	0.8	0.3
9. Northern California	0	0.7	0.2	0	1.7	0.4
10. Washington-Oregon	0	2.4	1.2	0	4.5	2.4
11. Lower Cook Inlet	0.5	4.7	1.4	1.0	14.0	4.1
12. Gulf of Alaska	0	0.2	0.04	0	0.5	0.1
13. Southern Aleutian Arc	0	2.4	0.7	0	5.3	1.6
14. Bristol Bay Basin	0	7.0	2.2	0	15.0	5.1
15. Bering Sea						
16. Chukchi Sea	0	14.5	6.4	0	38.8	17.5
17. Beaufort Sea	0	7.6	3.3	0	19.3	8.2
Total OCS			26.14			98.4

Source: USGS estimates.

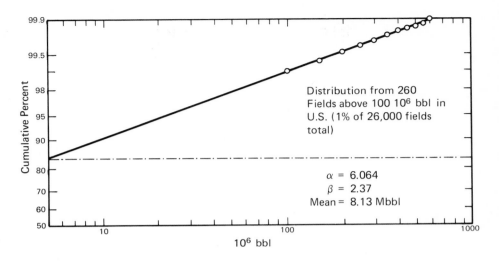

Source: Arthur D. Little, Inc., estimates.

Figure 3-1. Lognormal Distribution of United States Oil Field Sizes.

United States. Since fields smaller than 5 million barrels (oil equivalent) in all cases may not be considered commercial on the OCS, a truncated distribution has been used for the different OCS areas. Only the top 15 percent of the possible field sizes shown in Figure 3-1 will be developed on the OCS since all smaller field sizes are below the minimum economic field size under present cost/price relationships.

Fill Factor Distribution

Oil and gas fields have different fill factors in terms of the average number of recoverable barrels of oil per acre or average number of recoverable cubic feet of gas per acre. In the absence of knowledge of the specific fill factors which may be expected in a particular unexplored OCS area, the United States average fill factor distribution for giant fields has been selected as a best estimate of the distribution of the fill factor of OCS commercial fields. This distribution is presented on log-normal probability paper in Figure 3-2. The mean of this distribution is 56,750 barrels per acre, and it has a standard deviation of the log-normal distribution of 1.344 under the assumption that it is log-normally distributed.

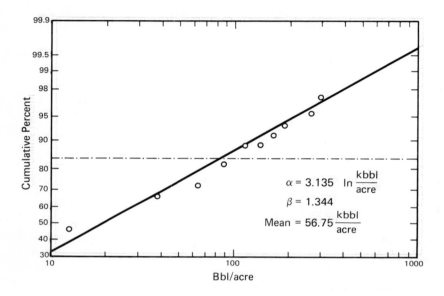

Source: Arthur D. Little, Inc., estimates.

Figure 3-2. Fill Distribution.

Structure Size Distribution

The distribution of structure sizes (in acres) is not publicly known for the larger fields of the unexplored OCS areas. An average structure size distribution has been derived from the United States average field size distribution (Figure 3-1) to serve as the basis for the present analysis and the United States average fill factor distribution (Figure 3-2) under the assumption that both distributions are log-normal. The resulting log-normal distribution is shown in Figure 3-3. The imputed structure size distribution has a mean of 31.2 acres, and the standard deviation of the log-normal distribution is 1.013. The particular distributions for the structural traps which have been used for the individual areas are given in Table 3-5. They have been based on the minimum economic field sizes as established for the different areas in this analysis.

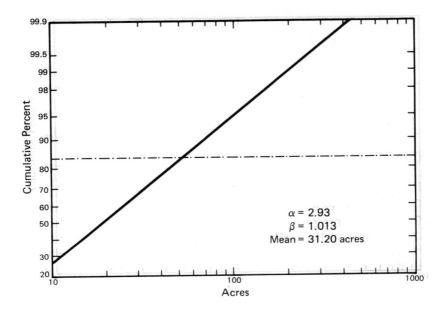

Source: Arthur D. Little, Inc., estimates

Figure 3-3. United States Average Structure Size Distribution.

Cost Data

All costs in this study are presented in 1975 dollars except where otherwise specified. The cost data base was developed from interviews with independent individuals and company representatives both in the United States and abroad. In addition, a literature survey was performed. The resulting data base contains estimates of all investment and operating costs from early seismic exploration activity up to delivery of offshore oil and gas at shore-based receiving facilities.

To allow for the wide range of different conditions which can be expected in the 17 different OCS areas, it was necessary to develop the costs at a level of detail where changes in these costs, because of contextual changes, could be allowed for properly. Consequently, this required the data items to be parameterized based upon our understanding of the engineering considerations and concepts on which the present-day offshore technology is based.

Apart from Upper Cook Inlet, none of the OCS areas of Alaska has had any offshore field development. The special circumstances, such as the extreme cold and harsh weather conditions, and other hazards not yet encountered in known areas, such as floating icebergs or moving ice fields, will require new platform designs and improved field development technology. The costs of this

Table 3-5
Size Distribution of Structural Traps as Used for Area Simulations[a] (*in square miles of surface area*)

Area Names	Cumulative Percentiles								
	0	1	5	25	50	75	95	99	100
1. Atlantic, Gulf of Mexico, Pacific	0.14	0.58	1.10	2.67	5.09	9.36	23.40	45.43	413.0
2. Alaska Offshore[b]	0.69	2.42	3.85	7.85	12.94	20.65	41.30	68.87	413.0
3. Beaufort Sea	0.96	4.54	6.88	12.66	19.27	28.91	55.07	82.60	413.0

Source: Arthur D. Little, Inc., estimates

[a]Assuming a minimum economic field size of 5 million barrels for the Gulf of Mexico, the Atlantic, and the Pacific, 15 million barrels for all areas south and west off the coast of Alaska, and 50 million barrels for the Beaufort Sea.

[b]Gulf of Alaska, Lower Cook Inlet, Bristol Bay, Bering Sea, and Chukchi Sea.

improved technology, as used in this study, could be estimated only by extrapolation of the costs of known technology as applied in areas with harsh conditions such as the North Sea. For this reason, the estimates which are presented here should only be taken for what they pretend to be—*educated guesses* of what it may cost to explore for and develop oil and gas fields in these new unknown areas.

Exploration costs are broadly defined as all costs incurred before the actual discovery of commercial oil or gas in a field. *Development costs* are all costs incurred to delineate a field and to install equipment and facilities necessary for production of that field including any transportation facilities and receiving terminals required to bring the oil and/or gas onshore. *Operating costs* (or *production costs*) are costs directly related to the production and transportation to shore of the oil and gas.

Only 7 of the 17 OCS areas, which have been analyzed in terms of their relative economics, have thus far seen actual exploration, development, and production activities. These are areas in the Gulf of Mexico and offshore southern California. The economics of exploration and development ventures of these areas are not directly applicable to the 12 other areas because of differences in weather conditions and in distances to major supply centers for oil drilling and for oil producing equipment. Several of these areas, such as the Gulf of Alaska, will require technology which, thus far, has not yet been used offshore the United States.

We believe that the technology developed over the past 6 to 7 years to find and produce oil and gas fields in the northern part of the North Sea will be applicable to most of the frontier areas which the Bureau of Land Management intends to open up for oil companies through lease sales over the next 3 years. Therefore, we have analyzed the technical costs which the oil industry have experienced while operating in conditions typical for the North Sea and the Gulf of Mexico, respectively, and these areas have been used as the two benchmark areas against which costs for the other frontier areas on the outer continental shelf were measured.

Generally speaking, environmental conditions as they are encountered in the Gulf of Mexico and offshore southern California can be considered the least severe for the United States. This, combined with the fact that construction sites and supply centers for equipment are all located very close to these areas, renders the Gulf of Mexico and offshore southern California the least costly in terms of unit exploration and unit development activities. Compared to the rest of the United States, the Gulf of Mexico has relatively small field sizes and relatively low well productivities, which have resulted in fairly high costs per unit produced or per well drilled, despite the low overall costs.

The exploration and development costs in the other OCS areas will generally be higher than in the Gulf of Mexico and the area offshore southern California. On a comparative basis, they will increase going to the north along the Atlantic and Pacific Coasts, gradually approaching northern North Sea costs, since the more severe weather conditions which prevail in the northern parts of the Atlantic and the Pacific are similar to North Sea conditions. The general expectation is that the Gulf of Alaska will require even higher exploration and development costs because of earthquake dangers in addition to the severe weather conditions and the short working season.

Seismic conditions, such as earthquakes, are not a problem in the other Alaska offshore areas. But other problems, like the occurrence of icebergs and the question of how to prevent collisions between floating icebergs and fixed structures, have not yet been resolved.

The timing of investments and other expenditures for development and production of a field are very important when assessing the overall profitability of a field. Estimates have been made of the average durations for the different areas for finding and developing different-sized fields under different circumstances. As for unit exploration and development costs, the duration of development activities will increase northward from the Gulf of Mexico area and the offshore southern California area with their mild weather conditions and their proximity to supply centers to the north along the Pacific Coast. The duration of development activities in the northern areas will be close to or longer than the lead times experienced in the northern North Sea and, as such, will increase the economic cost per unit produced.

Exploration and Appraisal

Exploration comprises all activities which companies undertake before they determine if commercial oil and gas are present in a certain area. In the United States, these activities can be broken down in two categories:

1. Prelease sale exploration activities consisting of magnetic,[a] graphimetric,[b] and seismic[c] surveys
2. Postlease sale exploration activities consisting of exploratory and appraisal drilling and more detailed seismic surveys

Companies buy leases, through a cash bonus bidding system, for the rights to explore for and develop oil and gas on tracts which generally have a size of 3 square miles, or 5760 acres.

---

[a]Measures changes in the earth's magnetic field occasioned by discontinuities in the earth's crust.

[b]Measures changes in the earth's gravity force.

[c]Measures the reflection of sound waves.

Seismic surveys are by far the most important of the three prelease sale type of surveys mentioned in providing companies with the first information about the type and size of structures underlying OCS areas for which BLM has announced a particular lease sale. The cost of these surveys is usually shared among several companies, which thus obtain the same basic data about the area—information which they interpret individually. This information gives the companies some indication about the possible location of oil and/or gas trapped in what are usually called structural traps. Table 3-6 shows an index of unit acquisition and interpretation costs for geophysical surveys of the different areas offshore the United States. Costs are listed relative to the benchmark area of the Gulf of Mexico (index 100).

The exploratory drilling is performed from a platform with legs which are adjustable in height, a *jack-up*, or from a floating platform, a *semisubmersible*, or from a specially equipped *drill ship*, depending on the particular conditions in the areas. Drill ships and semisubmersibles are generally used in waters deeper than 200 to 250 feet; jack-ups are reserved for shallower waters.

Construction costs of a jack-up rig in terms of 1975 dollars range between $20 and $30 million, depending on the particular area in which the rig will operate. When contracted by an oil company, the daily contract costs for the rig alone are between $20,000 and $30,000. Additional costs are incurred for supporting services such as supply boats, which can cost from $1200 per day in

Table 3-6
An Index of Marine Geophysical Survey Costs per
Line Mile for Acquisition and for Processing, 1975

Acquisition	
Gulf of Mexico (benchmark area)	100
Atlantic Coast	127
Pacific Coast	127
Gulf of Alaska	132
Chukchi Sea	127
Bering Sea	132
Beaufort Sea[a]	1136
Processing and Interpretation	
New Areas	95
Established Areas[b]	130

Source: Arthur D. Little, Inc., estimates.

[a]Beaufort Sea costs are assumed to be the same as average Canada land costs because surveys on ice tend to be more like land surveys than like sea surveys.

[b]Interpretation of data from established areas requires more of an effort because the large and obvious structures have already been explored.

the Gulf of Mexico to $4000 per day in Cook Inlet in Alaska. These costs, together with estimates of other costs such as casing and cementing costs, logging survey costs, drilling mud costs, helicopter costs, and mobilization and demobilization costs of the rig, are shown in Table 3-7.

When drilling in deeper waters, companies will contract a semisubmersible or a drill ship which is capable of drilling down to 25,000 feet in water depths of over 1000 feet. A large semisubmersible will cost $40 million to $50 million to construct and equip, which implies that an oil company will have to pay $40,000 to $50,000 per day[2] to the drilling contractor. Costs for other supporting and special services and for raw materials, shown in Table 3-7, can be another $20,000 to $30,000 for a typical well drilled in the North Sea. Since as many as 120 days can be required to drill an exploratory well (depending upon well depth), the cost of an exploration well in the North Sea can be as high as $9 million.

The number of days required to drill an exploration well is dependent not only on the well depth and the particular formations which have to be drilled

Table 3-7
Cost Breakdown for a Hypothetical 10,000-Foot Well, Northern North Sea Drilled by a Contractor Rig, Mid-1974

Activity	Cost Units	Cost ($)	Percent of Total
Preparation	Mobilization/Demobilization	147,000	3.9
	Site Preparation	25,000	0.7
	Transport Rig Move	84,000	0.9
Drilling	Contract Payments	2,000,000	53.1
Installation	Drilling Materials	11,000	0.3
Running	Fuel	85,000	2.3
Costs	Salaries	30,000	0.8
	Maintenance	23,000	0.6
Drilling	Mud	154,000	4.1
Materials	Bits and Coreheads	48,000	1.3
	Casing	245,000	6.5
	Cementing	41,000	1.1
Evaluation	Logging	135,000	3.6
	Intermediate Testing	15,000	0.4
	Miscellaneous Evaluation	30,000	0.8
Transport	Sea	545,000	14.5
	Air	124,000	3.3
	Overhead	68,000	1.8
	Total	3,760,000	100

Source: Society of Petroleum Engineers of AIME, Paper No. SPE 5266, "Drilling Costs," P. B. Jenkins and A. L. Crawford, 1975. Reprinted with permission.

through, but also on the prevailing weather conditions. This is illustrated by Figure 3–4, where the relation between average sea states and percent down time per month and cost per foot for wells drilled in the North Sea is shown. It can be seen that the exploration costs for the same type of well in the same area can fluctuate between $1 million and $9 million depending upon whether the particular well is drilled during summer or during winter.

The number of exploratory wells required to fully explore a given tract of 3 square miles can range from one, in the case where a very large and simple structure underlies the particular tract, to up to three or four, in the case of a more complicated geology as it exists in the Gulf of Mexico, for instance.

The two factors of wide variation in individual well costs and disparity in the number of wells required to fully explore a tract render it impossible to determine the precise cost for exploration of tracts in those OCS areas where drilling has not yet taken place.

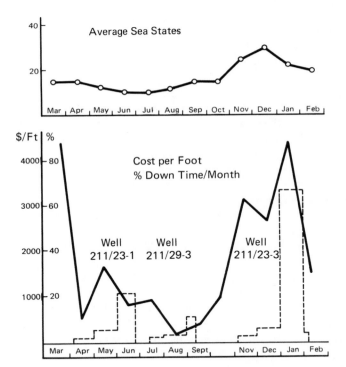

Source: Society of Petroleum Engineers of AIME, Paper No. SPE 5266, "Drilling Costs," P. B. Jenkins and A. L. Crawford, 1975. Reprinted with permission.

Figure 3–4. Relationship Between Sea State Cost Per Foot and Percentage Down Time (Based on data from the Sedco 135F).

If a discovery is made and oil and/or gas are found in commercial quantities, further drilling is required with the help of the exploration rig to further delineate the field. This field delineation, or appraisal drilling, can require another three to six wells depending on the complexity of the geology where the field has been found.

Table 3-8 shows the range of total exploration and appraisal drilling costs which a company may have to incur in the 17 different OCS areas to fully explore and appraise a tract of 5760 acres.

Development

The development of an offshore oil or gas field requires:

1. the construction and installation of a production platform
2. the manufacture and installation of production equipment
3. drilling of producing wells and wells used for injection of water and/or gas
4. the installation of facilities which enable transportation of the oil and/or gas to an onshore terminal

Oil and gas require treatment after they are produced and before they can be moved by pipeline to onshore terminals. A combination of environmental and economic considerations necessitates that the treatment be done on-site. This treatment consists mainly of separation of water and hydrocarbons since "formation" water is usually produced along with the oil or gas.

All the equipment required for treatment of the produced fluids, together with other types of equipment such as cranes, living quarters, a power plant, compressors, a helicopter landing deck, etc., is located on an artificial inland or platform which is standing on the sea bottom or is floating right above the particular oil or gas field.

The weight of the entire equipment and facilities package may total to 0.2 ton for every barrel of oil produced per day at peak capacity. For gas production platforms, the total weight is approximately 0.1 ton for every 10,000 cubic feet per day of peak capacity.

Currently, the majority of production wells are drilled from fixed platforms. The platform provides a stable basis from which these wells can be drilled and completed by using deviation drilling techniques from which areas in the reservoir, generally at depths between 5000 and 15,000 feet, can be reached as far out as 1 to 3 miles, measured from the vertical down from the platform.

The costs of fixed platforms increase exponentially with increasing water depth and increasing severity in weather conditions. Therefore, a strong economic incentive exists to look for alternate ways to develop the oil and gas fields which lay under deep waters or which have severe weather conditions. In the following

Table 3–8
Range of Total Exploratory and Appraisal Drilling Costs per Tract of 5760 Acres (1975 dollars)

Areas	Variable Costs ($1000/day)	Fixed Costs ($1 million per well)	Number of Days (Days/well)	Cost per Well ($1 million per well)	Wells per Tract	Cost per Tract ($1 million)
Atlantic Coast (1, 2, 3)	25–35	0.6–0.7	20–100	1.1–4.2	1–6	1.1–25.2
Gulf of Mexico (4, 5, 6)	25–35	0.5–0.6	20–100	1.0–4.1	1–6	1.0–24.6
California (7, 8, 9)	25–35	0.5–0.6	20–100	1.0–4.1	1–6	1.0–24.6
Oregon and Washington (10)	25–35	0.6–0.7	20–100	1.1–4.2	1–6	1.1–25.2
Alaska, South (11, 12, 13)	50–75	0.8–0.9	30–120	2.3–9.9	1–6	2.3–59.4
Alaska, East (14, 15)	40–55	1.0–1.2	30–120	2.2–7.7	1–6	2.2–46.2
Alaska, Northeast (16)	40–55	1.1–1.3	30–120	2.3–7.9	1–6	2.3–47.4
Alaska, North (17)	40–55	1.2–1.4	30–120	2.4–8.0	1–6	2.4–48.0

Source: Arthur D. Little, Inc., estimates.

section, the technological costs of the more conventional type of field development using wells drilled and completed from fixed platforms are compared with the costs of a newer alternative using subsea completion technology and floating platforms. This latter type of development by now has reached the prototype stage and is being used in the development of several North Sea oil fields. The success of this technology will set the rate at which subsea completion technology can be expected to be used in offshore oil and gas development in the frontier areas of the United States. Technology for transporting oil and gas from the field to onshore receiving terminals has also undergone significant changes; increasing water depth, more severe weather conditions, and longer distances to shore have resulted in increases in costs for bringing the oil and gas onshore.

For gas, cost reduction for transporting the gas onshore is limited to improvements in pipe laying and burying techniques. Liquefaction at the OCS field site enables transportation of the gas by tankers but may be too costly relative to present-day prices of $.50 to $.60 per thousand cubic feet.[d] Even floating offshore LNG plants will probably not be able to produce gas for distribution at current United States market prices. Also, the floating LNG plants currently under construction are intended for the relatively calm environment of the Arabian Gulf and the Java Sea, where LNG tankers can moor alongside.

For oil, tanker transportation to shore has been shown to be an attractive alternative compared to pipeline. The costs for these two alternatives are presented below as a function of the maximum capacity of the transportation system and the distance to shore.

Platform. Figure 3–5 shows the various alternative platform constructions which industry is presently using or testing offshore.

The *conventional steel jacket* is the original type of platform of the industry. In the Gulf of Mexico, steel jackets have been used for over 20 years in water depths of up to 350 feet while in the North Sea, steel platforms in water depths of up to 450 feet have been installed.

Concrete platforms were introduced in the North Sea to minimize costs and delivery times. Concrete platforms, however, now cost at least as much as steel platforms for the same water depth and weather conditions, but they can include oil storage capacity of up to 1 million barrels for a small additional cost (about 5 percent). In addition, they can accommodate production equipment for capacities of over 200,000 barrels per day, which is beyond the upper limit for steel platform designs for the North Sea. They do, however, require a deep water inlet for construction.

---

[d]When this was being written, wellhead prices of gas were still at a maximum of $.52 per thousand cubic feet. Since then the FPC released that wellhead prices for new gas may go up to $1.42 per thousand cubic feet. The final approval of these new higher wellhead prices is subject to court decisions in suits brought by groups which oppose the price hike.

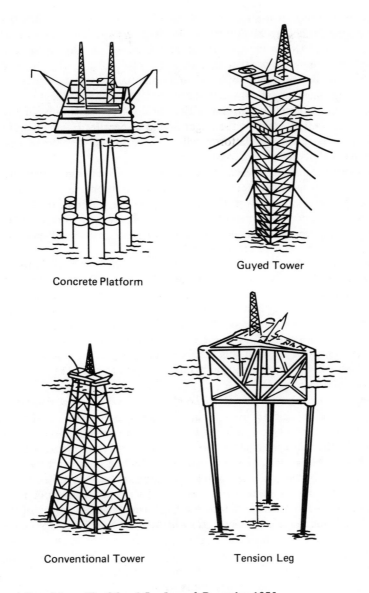

Concrete Platform

Guyed Tower

Conventional Tower

Tension Leg

Source: Adapted from *The Oil and Gas Journal*, December 1975.

Figure 3–5. Alternative Fixed-platform Constructions for Offshore Production of Oil and Gas.

It is expected that the size of concrete platforms (150,000 to 300,000 tons and requiring 70,000 to 80,000 hp for towing) makes it difficult to tow them over the Atlantic Ocean for use off the East Coast of the United States. Deep-water inlets suitable for construction are not easily available. The Canadian East Coast (e.g., the Bay of Fundy) offers several potential construction sites for these platforms, although large tidal variations may prove to be a problem.

The only potential construction site for concrete platforms on the West Coast of the United States is the Puget Sound at Seattle. It will, therefore, be more likely that concrete platforms will be used in the development of areas off the West Coast and off the coast of Alaska.

In addition to the all-steel and all-concrete platform designs, there are other designs which combine concrete and steel, each with its claim on cost advantages over the all-concrete and all-steel designs. However, as yet, none of these designs has been tested under field conditions, which precludes any forecast about their cost. It can be expected, though, that some of these designs will eventually be used in the development of offshore fields.

The *guyed tower* concept is now undergoing a small-scale test in the Gulf of Mexico to test this platform type for development of fields in the very deep waters in the Santa Barbara area off the coast of southern California. This tower design saves on steel requirements for the tower by dissipating part of the wave and wind energy exerted on the platform through the guidelines rather than through the structure itself, as in the case of a conventional steel platform. The current experimental stage of this platform concept precludes an assessment of what its cost will be. Exxon reportedly intends to use this type of tower in water depths of 1500 feet.

The *floating platform* used in combination with *subsea completions* is another alternative which is used in the development of certain smaller fields in the North Sea. The obvious advantage is that the cost of fabrication and installation of the platform is much less dependent on the particular water depth in which it is being used. The sensitivity of a floating platform to wave movement requires that the wells be completed on the sea bottom rather than on the platform itself. The production from individual wells can then be combined by a subsea manifold and delivered into the treatment facilities on top of the platform through one single pipeline or riser which, under severe weather conditions, can be disconnected between the floating platform and the sea bottom. In spite of the limited experience with this type of development system for an oil field, enough cost information is available to make a tentative comparison with the other more conventional fixed platform systems.

Platform Capacity and Well Productivity. The deck load of a platform is a function of the maximum design capacity of the platform production equipment. There can be large variations in type and size of equipment used to treat a given amount of produced fluids, depending on whether these fluids are dry gas or gas with condensate, heavy crude oil with only a trace of associated gas or light crude with a relatively high gas-oil ratio. In addition, it is possible that the particular type of reservoir will require pressure maintenance through water injection and/or gas reinjection, which then will result in additional equipment requirements.

The maximum capacity which existing platforms can accommodate is constrained, on the low end, by the maximum number of production wells (normally assumed to be 40); on the high end it is constrained by the maximum platform size which can be constructed. The largest steel platforms that have been constructed for the North Sea are capable of accommodating production equipment for up to 150,000 barrels per day of crude oil and up to 200 million standard cubic feet per day of gas in addition to water injection equipment for up to 300,000 barrels per day. Concrete platforms now under construction for the Statfjord field will be able to handle up to 300,000 barrels per day of crude oil and to treat and reinject up to 0.5 million cubic feet of gas per day in addition to reinjection of up to 400,000 barrels per day of water.

The number of wells that can be drilled from a given platform will depend upon:

1. the reservoir characteristics of the particular oil or gas reservoir, such as the porosity, connate water[e] saturation, permeability, and type of drive mechanism
2. the height of the produced oil or gas column in the reservoir
3. the average depth of the reservoir

As shown in Table 3-8, the maximum area which can be produced from one fixed platform is dependent on the depth of the reservoir. Under the assumption that a deviated well can be drilled to an angle of up to 50° with the vertical, the maximum number of acres to be produced from a single platform for an oil field, typically found at a depth of between 5000 to 10,000 feet, can range from 2000 to 8000; for a gas field, which will typically be found between 10,000 and 15,000 feet, this can range from 8000 to 18,000 acres.

The number of wells which have to be drilled to produce the oil and/or gas contained in the area shown in Table 3-9 will depend upon the well spacing, that is, upon the number of acres of reservoir that can be produced by one well. As mentioned earlier, this well spacing is mainly a function of the type of reservoir fluid produced, oil or gas, and of the reservoir characteristics such as the connate water saturation, porosity, permeability, and the driving mechanism. It is beyond the scope of this analysis to show how well spacing can vary as a

---

[e]The porous spaces in most reservoir rocks were originally filled with water which was then replaced by oil and/or gas, leaving only a film of water on the rock surface—the connate water.

Table 3-9
Maximum Size of Area Which Can Be Produced with Deviated Wells Drilled from a Single Platform[a]

Depth of Reservoir[b] (ft)	Maximum Size of Area Which Can Be Produced with a Single Platform (acres)[c]
5000	2000
7500	4500
10,000	8000
12,500	12,500
15,000	18,000

Source: Arthur D. Little, Inc., estimates.

[a]Assuming a maximum angle of deviation with the vertical of 50°.

[b]The range of 5000 to 10,000 feet is representative for oil reservoirs, while the range of 10,000 to 15,000 feet is more typical for gas reservoirs.

[c]The maximum size of a tract offshore the United States is 5760 acres, or 3 square miles.

function of each of these parameters. Therefore, well productivity and recoverable reserves per acre were used as two composite parameters with which well spacing will vary.

Using what can be considered to be a typical production profile for an oil well with a producing plateau, at peak capacity, of about 5 years, followed by a period of decline of 15 years at 15 percent per annum, it was calculated what the well spacing would have to be at different values of recoverable reserves ranging from 10,000 to 200,000 stock tank barrels per acre. The results are shown in Figure 3-6, where the range of values found in the Gulf of Mexico and the North Sea for well productivities and recoverable reserves are indicated by shaded areas.

In a similar fashion, recoverable reserves have been estimated for gas reservoirs ranging from 50,000 to 400,000 million standard cubic feet per acre and well productivities ranging from 10 to 80 million standard cubic feet per day. For gas, a typical production profile was considered with a peak production plateau of 10 years followed by a 30-year period of declining production, declining at an annual rate of 10 percent. Figure 3-7 shows the results with indications of typical values for gas well spacings in the Gulf of Mexico and in the North Sea.

It should be noted from Figures 3-6 and 3-7 that gas fields, in general, admit larger well spacings than oil fields. Gas fields have well spacings between 500 and 8000 acres per well; oil fields have more typical well spacings of 80 to 2000 acres per well.

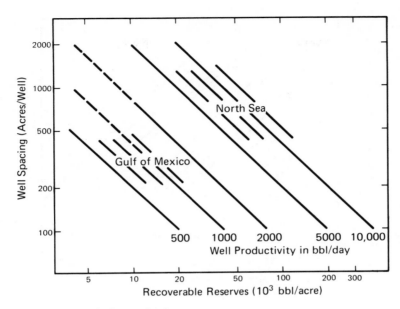

Source: Arthur D. Little, Inc., estimates.

Figure 3-6. Well Spacing as a Function of Recoverable Reserves per Acre (Oil) (Assuming a 5-year plateau followed by 15 years of decline at a rate of 15 percent per annum).

Combining the information on range of well productivities, recoverable reserves per acre, and area of reservoir to be produced at different depths from one single platform, one can estimate the range of platform sizes that will be required in the unexplored OCS areas. In Figure 3-8, it is shown that full use of economies of scale for fixed platforms by use of the maximum sized platforms which can currently be constructed is only possible under exceptionally fortunate circumstances where the oil reservoir is found at a depth of approximately 10,000 feet and with reservoir characteristics allowing very high well productivities of around 10,000 barrels per day from a thick reservoir with recoverable reserves of around 150,000 barrels per acre and total recoverable reserves of 500 million barrels.

Platform Fabrication Costs. The much more severe weather conditions in the North Sea exemplified by a design wave ranging from 90 to 100 feet, compared to the Gulf of Mexico where the design waves are more in the range of 60 to 70 feet, requires much heavier structures for the same water depths. This can be seen in Figure 3-9 where, for instance, in water depths of between 200 and

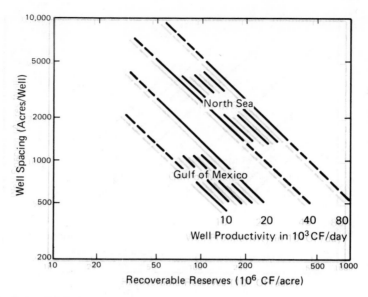

Source: Arthur D. Little, Inc., estimates.

Figure 3-7. Well Spacing as a Function of Recoverable Reserves per Acre for Gas (Assuming a 10-year plateau followed by 20 years of annual decline at a rate of 10 percent per annum).

400 feet, steel jackets in the Gulf of Mexico require 10 to 13 tons of steel per foot of water depth compared with steel jackets in the North Sea requiring 20 to 38 tons of steel per foot of water depth. For comparison purposes, weight estimates for a steel jacket strong enough to withstand weather conditions and earthquakes in the Gulf of Alaska are shown in the same figure.

Figure 3-10 shows actual and estimated construction costs of steel and concrete substructures in various offshore areas, relative to the cost of those structures in the benchmark area of the Gulf of Mexico. In interpreting Figure 3-10, it should be realized that the deck loads which are typically required for the Gulf of Mexico, as a rule, do not exceed 20,000 barrels of oil per day, or 200 million cubic feet of gas, whereas oil production capacities for platforms in the North Sea typically range between 100,000 to 200,000 barrels per day, or 5 to 10 times as large.

For comparison purposes, we have also included in Figure 3-10 the range of costs for the fabrication of substructures of jack-ups and semisubmersibles. Semisubmersibles are used in the North Sea in the development of two small fields, in conjunction with subsea completions. In these cases, it has been found to be more economically attractive to develop the relatively small fields of 50

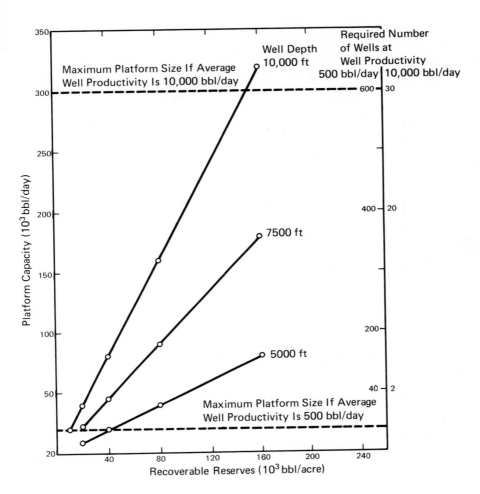

Source: Arthur D. Little, Inc., estimates.

Figure 3-8. Oil Maximum Required Platform Capacity and Required Number of Wells as a Function of Recoverable Reserves per Acre for Different Well Depths

to 100 million barrels of recoverable oil with four or five wells drilled from a semisubmersible and completed on the sea bottom, and to produce those wells into separation and treatment equipment located on top of a converted semisubmersible.

There is strong indication that in high-cost areas such as the North Sea, larger fields initially may be developed in this manner as well. In this case,

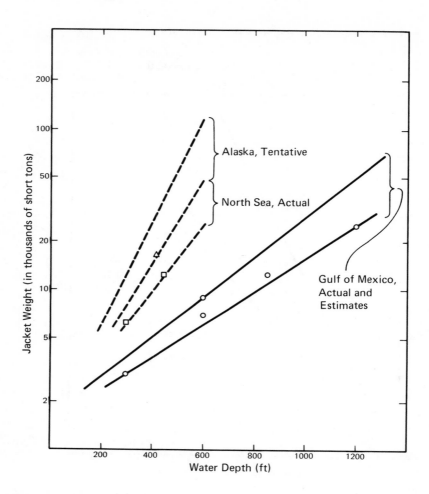

Source: Arthur D. Little, Inc., estimates.

Figure 3-9. Steel Jacket Weights versus Water Depth.

subsea completion would be an intermediate solution to obtain more information about the actual reservoir characteristics and to obtain positive cash flow while waiting for the construction of the large fixed platform which usually takes 2 to 3 years.

Cost data available for substructures which have been constructed or are under construction for use in the North Sea indicate that the weather load factor is dominant in the overall design. An analysis of platform construction costs as a function of capacity show that these fabrication costs are relatively

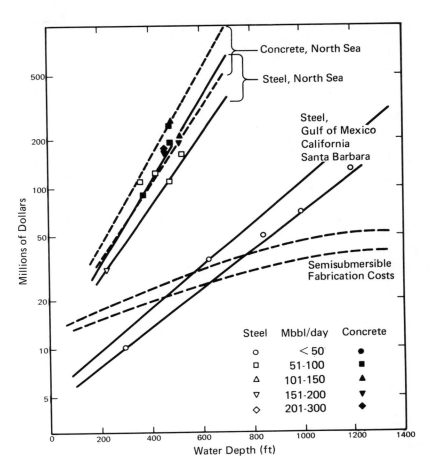

Source: Arthur D. Little, Inc., estimates.

Figure 3-10. Platform Fabrication Costs (1975 $).

insensitive to capacity over the range from 50,000 to 300,000 barrels per day. However, it must be realized that most designs for which cost estimates are shown in Figure 3-11 are still in the prototype stage and that they most probably will undergo significant cost-reducing improvements. This can be exemplified by the three platforms which were constructed for the Forties field, where as a result of design optimization, more than 25 percent steel was saved for the construction of the third platform in spite of the fact that the platform was designed for 450 feet while the first platform had been designed for 415-foot water depth. Whether platform costs in the future will be reduced by

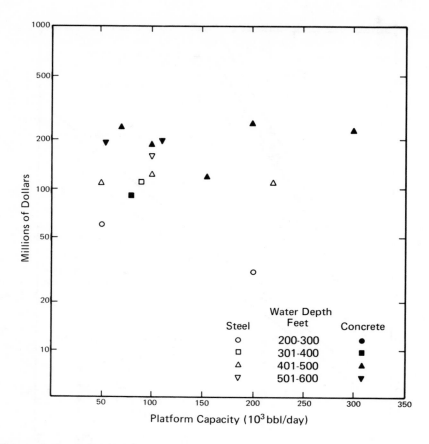

Source: Arthur D. Little, Inc., estimates.

Figure 3–11. Platform Fabrication Cost as a Function of Capacity.

such engineering optimizations as compared to costs in 1975 will depend on how the cost of labor and the cost of materials change and on the supply and demand for platforms in general over the next few years.

Based on the analysis of platform construction costs for areas presently under development, supplemented with the results of discussions with industrial sources, the range of construction costs for differently-sized platforms for the various areas on the OCS has been estimated. For this purpose, the 17 areas were classified into four regions, as follows:

1. the Gulf of Mexico and the Pacific Coast: OCS areas 4, 5, 6, 7, 8, 9, and 10[f]

---

[f]BLM OCS area classification.

2. the Atlantic Coast: OCS areas 1, 2, and 3
3. the eastern Alaskan Coast: OCS areas 14, 15, and 16; and the southern Alaskan Coast: OCS areas 11, 12, and 13
4. the Beaufort Sea: OCS area 17

As shown in Figure 3-12, weather conditions are quite different in the Gulf of Mexico as compared with the Pacific Coast. Maximum recurring wave height and wind speed in the Gulf of Mexico are considerably higher than anywhere along the United States part of the Pacific Coast. In spite of this difference in weather conditions, platform construction costs for offshore southern California are comparable with those for the Gulf of Mexico since platforms constructed for use in the waters offshore southern California must be able to withstand earthquakes. This results in the use of platform structures quite comparable to the structures used in the Gulf of Mexico with its more severe weather conditions.

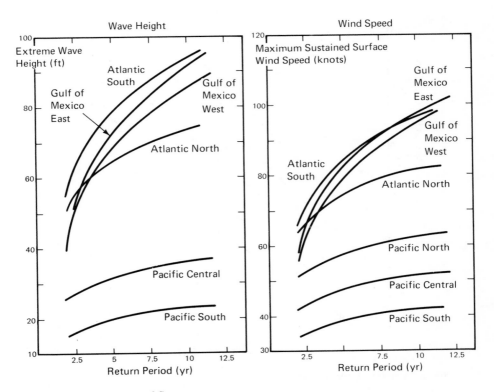

Source: U. S. Department of Commerce.

Figure 3-12. Recurring Maximum Wave Heights and Wind Speeds of the East Coast and of the West Coast.

Figure 3-13 shows how estimated platform construction costs change as a function of water depth in the different areas. Figure 3-14 shows how platform costs are expected to change as a function of the required treatment capacity for oil and gas. Based on an inspection of the data for gas platforms, it was concluded that the same platform size will be required for 10 MCF of gas per day as is needed for 1 barrel of oil per day.

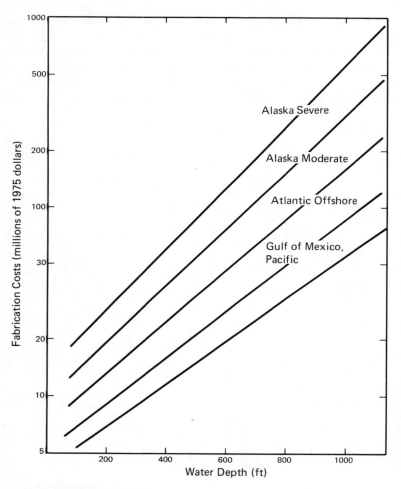

Source: Arthur D. Little, Inc., estimates.

Figure 3-13. Platform Fabrication Costs as a Function of Water Depth for Different Areas; Capacity 20 10^3 bbl/day, 200 10^6 CF/day (in 1975 dollars).

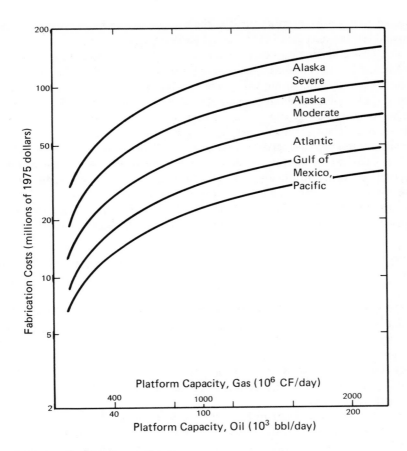

Source: Arthur D. Little, Inc., estimates.

Figure 3–14. Platform Fabrication Costs as a Function of Capacity for Different Areas; 300 ft of Water (in 1975 dollars).

The expectation is that platform structures to be used offshore northern California and offshore Washington or Oregon will be comparable to the structures used in southern California. Earthquake danger offshore Washington and Oregon is considerably less than it is in southern California; but weather conditions are more severe, especially during the winter season, and it can be expected that platforms will have to accommodate larger stocks of drilling material to enable continuation of the development drilling during periods of bad weather.

The same argument—that a high chance for long periods of bad weather interfering with the supply of drilling materials will require design for heavier loads—applies also to the Atlantic Coast areas. Hence, since weather conditions in these areas are more severe than those in the Gulf of Mexico, it can be expected that platform structures for these areas will be more expensive than comparable structures used in the Gulf of Mexico (see Figure 3–12).

Platform construction costs for the eastern Alaskan Coast are expected to be comparable with platform costs for the northern North Sea. The moving ice during the winter and spring will require strong and heavy structures, and the extreme cold will require the use of special high-grade steel. It has to be remembered, though, that it is not altogether certain that platforms will be used in these areas because of the danger of collision with icebergs.

The most expensive platform structures will be needed for the southern Alaskan Coast where platform structures should be able to survive earthquakes frequently occurring in the area, as well as weather conditions which are quoted to be even more severe than those in the northern North Sea area. If platform structures are used for field development in the Beaufort Sea, then they can be expected to cost at least as much as structures used in the Gulf of Alaska.

Platform Equipment Costs. The equipment on a production platform can consist of:

1. Oil treatment equipment, such as separators, to remove the formation water and the gas produced with the oil; oil metering and oil storage facilities and pumps to move the oil to shore through the pipeline or to a single-point mooring buoy for transportation by tanker.
2. If substantial volumes of nonassociated gas are produced, a natural liquid gas plant to remove the condensate and gas dehydrators to remove the water from the gas.
3. A water treatment plant to bring the concentration of hydrocarbons in the produced formation water down to a prespecified level and, in the case of a large platform, a sewage treatment plant, both of which are required to comply with pollution control standards.
4. A power plant and, in the case of gas production which is not flared, gas compressors.
5. In the case of water injection for pressure maintenance purposes, saltwater treatment facilities and injection pumps.
6. Living quarters for personnel.
7. A helicopter deck.
8. Hoisting and lifting equipment.
9. Fire-fighting and safety equipment.
10. Drilling equipment.

Some platforms also have a flare stack to burn gas that is not being reinjected or moved by pipeline. In cases where the volumes of gas to be flared are large, such a flare stack will be positioned at a safe distance from the platform on a separate small structure. This structure can be either a light jacket standing on the sea floor or a floating tower attached to the sea bottom by an articulated joint. In the Gulf of Mexico, it is quite common to find the production facilities on a separate platform next to the platform from which the wells have been drilled and completed.

Table 3-10 shows the cost breakdown for a large platform in the North Sea accommodating 32 wells, 12 of which are used for gas reinjection and water injection. The construction equipment has been sized to handle up to 125,000 barrels per day of crude oil, 200 million standard cubic feet of gas per day, and 200,000 barrels of seawater per day, respectively. Figure 3-15 shows actual and estimated costs, if still under construction, for total packages of production equipment over a wide range of capacities in the Gulf of Mexico and in the North Sea. It is evident that these costs, for a given production capacity, can vary up to 30 percent. Production equipment costs for concrete platforms are considerably higher than production equipment costs for steel platforms because of the difference in construction methods. Production equipment and other facilities for steel platforms are constructed as modules, each module weighing not more than 1000 tons so that it can be lifted and fitted into its place on top of the platform after having been transported to the platform location when the platform structure was installed. In the case of a concrete platform, the production equipment and other facilities are constructed as one single package, which is put on top of the concrete substructure, at the construction site, before the complete platform is being towed to its field location.

The range over which production equipment costs for a steel platform vary, for a given capacity, has been divided into four smaller ranges to allow situational cost differentials:

Table 3-10
Typical North Sea Platform Equipment Costs
(Capacity 125 10^3 bbl/day Oil, 200
10^6 CF/day Gas)

	Percent Breakdown of Total Costs
Oil Production Equipment	25
Natural Liquid Gas Plant	20
Water Injection Plant	5
Power and Switch Gear	20
Living Quarters and Helideck	10
Hoisting and Lifting Equipment	5
Fire Fighting and Safety	5
Drilling Equipment	5
Miscellaneous[a]	5
Total	100

Source: Arthur D. Little, Inc., estimates.

[a]Mainly radio tower and gas flare.

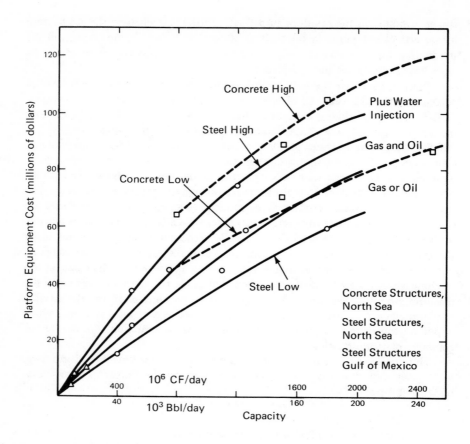

Source: Arthur D. Little, Inc., estimates.

Figure 3-15. Costs of Platform Production Equipment and Facilities (1975 dollars).

1. The production consists mainly of oil.
2. Oil will be produced with a substantial amount of gas, which will be reinjected and/or transported to shore.
3. Mainly oil will be produced, and the reservoir pressure will be maintained through water injection.
4. Oil will be produced together with a substantial amount of gas which will be reinjected and/or transported to shore, and the reservoir pressure will be maintained through water injection.

This subdivision was based on a more detailed assessment of relative costs for equipment required for gas transportation and/or reinjection and water injection.

The largest part of these costs consists of pumps and gas compressors for reinjection and/or transportation purposes.

Platform Installation Costs. Platform installations can comprise 30 to 35 percent of the total platform costs in the case of steel platforms and 10 to 15 percent in the case of concrete platforms.

Steel jackets, such as the large ones used in the North Sea, will be towed out of the construction dock while floating on special flotation tanks, or will be loaded on a barge as in the case of the smaller jackets used in the Gulf of Mexico. The jacket will be towed to the installation site with the help of four to six ocean-going tugs, where the jacket will be put into an upright position while it is sinking to the bottom. A temporary deck will be installed on top of the jacket, to accommodate the pile-handling units which will drive the piles used to fix the jacket firmly in place. The temporary deck will be removed when the piling and grouting have been finished after which cap trusses will be installed followed by the deck modules weighing up to 1000 short tons each with the help of a large derrick barge. When this is finished, the mechanical and electrical hook-up can take place. This whole operation can take from 4 to 6 months, depending on weather conditions and unforeseen complications during the installation.

At least one derrick barge will be required on-site during the installation of the jacket and modules, and it costs between $70,000 and $150,000 per day, depending on size. An example of the relative costs of the different types of support equipment and activities during installation of a platform is given in Table 3-11 where a breakdown is shown of the installation cost for a 125,000-barrel-per-day platform in 450 feet of water in the North Sea.

It is assumed that the installation costs for a platform will be between 30 and 35 percent of total platform costs if the platform is installed offshore the East Coast, offshore the West Coast, or in the Gulf of Mexico. In these regions, suitable construction sites for steel platforms can be expected to be available within 500 miles from the areas where offshore oil or gas fields may be found. To allow for the greater distances to the different areas offshore Alaska, the transportation charges for the jacket and deck modules are calculated on a per-tonnage basis to increase the total installation costs accordingly.

Subsea Completions. Research and development work on subsea completion technology has been conducted since around 1964. Initially, the efforts were concentrated on the development of "wet Christmas trees" which would make possible the drilling and completion of wells in those parts of the oil or gas fields that could not be reached from the platform. The considerable increase in interest of the oil industry in development of offshore oil and gas fields and the availability of technology and engineering concepts developed for the space program resulted in the development of *dry* subsea completion systems. These

Table 3–11

A Breakdown of Platform Installation Costs for Jacket and Deck Sections for a Platform Accommodating a Production of 125 10^3 bbl/day of Oil and 200 10^6 CF/day of Gas in Water Depth of 450 Feet

	Percent of Total Costs
Temporary Work Decks[a]	3.0
Modify Barges	1.5
Derrick Barges	45.0
Cargo Barges	9.5
Tugs and Supply Vessels	10.5
Diving	7.0
Pile-handling Units	2.0
Grouting	0.5
Miscellaneous Installation	3.5
M&E Hook-up	16.5
NGL Installation	0.5
Storage and Handling	0.5
Total Installation Costs	100

Source: Arthur D. Little, Inc., estimates.

[a]Used for platform piling.

provide an atmospheric working environment around the Christmas tree on the sea bottom, precluding special training of oil production personnel for routine service and maintenance work on individual wells. Reportedly, Exxon is working on a remotely controlled "wet" system.

Figure 3–16 shows different elements of a typical "wet" subsea field development consisting of the subsea wellhead completions and a subsea manifold system which combines the production from various wells into one line. Through this line the production is delivered into treatment and separation equipment on top of a floating platform and from there into a tanker through a single-point mooring buoy. This type of system, which is now being used in a full-scale development of the Argyll field in the North Sea in water depth of up to 420 feet, is operational to water depths of up to 1200 feet. The development of a subsea production station, performing the functions now performed by the production station supported by the floating platform, will most probably take at least another 8 to 10 years, mainly because of difficulties to supply the power required to operate a production station on the bottom of the sea.

The economic incentive to use subsea completion systems, rather than wells which are drilled, completed, and serviced from a fixed platform, increases exponentially with increasing water depth. The advantages of subsea completion technology are many:

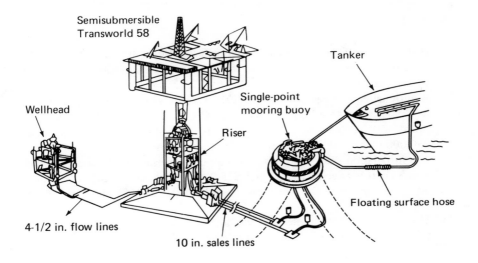

Source: Adapted from *The Oil and Gas Journal*, June 1975.

Figure 3-16. Subsea Completion System.

1. It may allow the production of those parts of the field which cannot be reached from a fixed platform but which are not large enough to justify the installation of another platform.

2. Subsea completions may be used to produce from appraisal wells at the initial stage in the development of a large field before the exact plan for the full development of the field is chosen. This would offer the combined advantage of an early positive cash flow and acquiring additional information about the reservoir characteristics.

3. Potentially, one or more platform structures may be saved in the full development of a large field. Only one platform could be used to separate and treat the produced formation fluids. This platform would not necessarily be positioned on top of the field, but could be installed in shallower waters close to shore.

4. A number of smaller fields in deep waters could be developed, using a floating platform as production station, when investment in a large fixed platform cannot be economically justified.

When the costs of the completion of a subsea well are being considered and compared to those for a well completed on top of a platform, the following increases in costs for conditions such as those in the Gulf of Mexico can be identified:

1. Higher well drilling costs because wells will have to be drilled using a jack-up or semisubmersible which will be more expensive than drilling wells from a fixed platform.
2. Additional equipment costs for the cellar[g] are estimated to be approximately $750,000 to $1 million for water depths up to 400 feet. For each additional 400 feet of water depth capability, the cost will increase between $50,000 to $100,000.
3. The installation costs for the subsea completion using a special service ship which costs between $16,000 and $25,000 per day in the Gulf of Mexico. The same service in the North Sea costs between $32,000 and $50,000 per day.
4. The cost of gathering lines and the cost of a subsea manifold add another $200,000 to $300,000 to the total subsea completion cost.

The incremental subsea completion costs for North Sea conditions are generally estimated to be between 1.5 and 2 times higher than those for the Gulf of Mexico. The total incremental costs for a subsea completion for water depths of up to 1200 feet in the Gulf of Mexico are estimated to range from $1.2 million to $1.8 million and for the North Sea to range from $2.1 million to $3.3 million per well. These estimates of incremental subsea completion costs are shown in Figure 3-17 together with the platform construction and installation costs per well, assuming that each platform would accommodate 20 producing wells. On the basis of this comparison, subsea completions appear to become economically attractive in the Gulf of Mexico in water depths between 400 and 650 feet and in the North Sea in water depths between 200 and 300 feet. This clearly illustrates that the more severe the weather conditions and the deeper the water, the more attractive subsea completion technology will become.

In the analysis of minimum economic field size, the extent to which application of subsea completion technology might help to make otherwise submarginal fields economically feasible projects has been shown.

Transportation of Production to Shore. There are basically two ways to transport oil or gas to shore. A submarine pipeline can be laid from the particular oil or gas field to a receiving terminal onshore, or tankers can be loaded on-site at the field which then transport the oil or gas to whatever receiving point might be most feasible.

---

[g]This is the capsule which accommodates valves and connections controlling the outflow of the oil and/or gas streams.

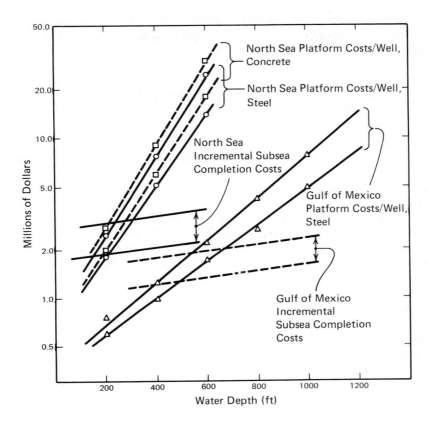

Source: Arthur D. Little, Inc., estimates.

Figure 3-17. Platform Fabrication and Installation Costs per Well Compared with Incremental Subsea Completion Costs.

For natural gas, on-site liquefaction would be required to allow transportation by special LNG tankers. Both the liquefaction plant and the tankers require large capital investments, which under current price conditions would not be economically justifiable offshore the United States. Also, loading LNG tankers on the high seas would require solutions to some very special technological problems, such as the design of a cryogenic flexible hose. Liquefied natural gas plants are now being constructed for use in the Persian Gulf and the Java Sea at the field site in offshore areas. The very calm waters in these areas greatly simplify the loading problems as compared with similar operations, for instance, in the Gulf of Mexico. Therefore, in the present analysis (of minimum economic field sizes) *it is assumed that, so far, shipment of natural gas in the form of LNG cannot be considered to be feasible.*

For oil, the alternative of shipment by tanker has been proved economically attractive under circumstances where economies of scale through use of large pipeline sizes to accommodate production of several fields cannot be used.

Pipeline Costs. Pipeline costs are broken down into material costs, pumping station costs, pipe-laying and pipe-burying costs, and costs for the shore approach. In general, it is expected that pipeline costs in dollars per mile will increase with increasing pipe diameter, weather severity, distance, and pipeline water depth.

Pipe-laying operations have to be suspended when wave heights exceed a certain level. First-generation lay barges can operate only when waves are less than 5 feet; third-generation barges can continue operating at waves of up to 15 feet.

The diameter of the pipeline is mainly a function of the maximum throughput expected and the maximum pressure at which the oil or gas will be pumped. Figures 3-18 and 3-19 show optimum line sizes and their costs for a given distance as a function of maximum daily throughputs for oil and gas lines, respectively.

In the case of gas lines, where more pumping stations are required to move the gas over the same distance as for oil, the economics of the optimum line size can be changed considerably by foregoing a pumping station and using a larger-diameter pipeline. The increases in material and laying costs of pipes are then offset by the decrease in pumping station costs for which an extra platform is also required.

Steel costs can range between $500 and $1500 per metric ton for the steel depending on the diameter of pipeline, the grade of steel used, and whether standard pipe is used. Wrapping with tar paper and coating with 0.5 to 3 inches of concrete can cost between $50 and $150 per metric ton. Lay barges usually require adaptations for the specific job which can cost up to $3 million for the third-generation barges; mobilization and demobilization of the barge which, depending on the distance over which the barge will have to be moved before it starts laying pipe, can be between $200,000 and $2 million; daily costs for a lay barge, depending on whether it is first, second, or third generation, will range between $70,000 and $170,000 per day; bury barges will cost between $60,000 and $100,000 per day.

The number of days that barges will be required to lay a pipe over a certain distance will be a function of: (1) the good weather laying rate which currently is between 1 and 2 miles per day for diameters of up to 36 inches in water depths of up to 500 feet; (2) the weather down-time factor which, in the North Sea, is close to 3, implying that the barge is waiting for good weather or picking up abandoned pipe 2 out of every 3 days; and (3) the material maintenance factor which is usually taken to range between 0.65 and 0.75. The latter figures imply that in spite of maintenance work during weather down time, 25 to 35

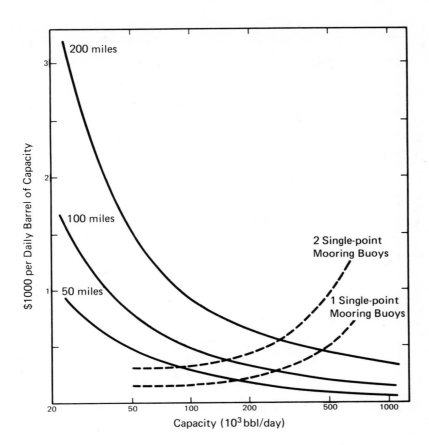

Source: Arthur D. Little, Inc., estimates.

Figure 3-18. Oil Pipeline Costs as a Function of Capacity.

percent of the good weather time still has to be used for maintenance activities which interfere with pipe laying.

Landfall or shore-approach costs are more complex to estimate because they are completely dependent on the shore conditions. Depending on whether the shore approach is a smooth, gently sloping, sandy beach or a rough, rocky coast with outcrops which require removal by underwater blasting, these costs can vary from $1 million to $12 million.

Figure 3-20 shows the range of line-mile costs for a 500,000-barrel-per-day line and a 50,000-barrel-per-day line when different assumptions are made for the distance-to-shore and the weather down-time factors. All four cost categories (materials, laying and burying, pumping stations, and shore approach) show a

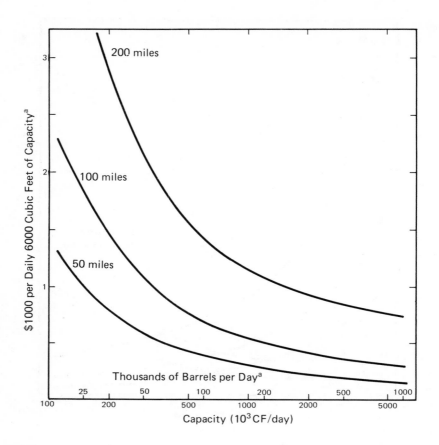

^{a}6000 cu ft/day = 1 bbl/day of crude oil equivalent.

Figure 3-19. Gas Pipeline Costs as a Function of Capacity (1975 dollars).

considerable range if we compare a low-cost, 200-mile line with a weather down-time factor of 1 to a high-cost, 25-mile pipeline with the weather down-time factor of 4. Total costs for the 500,000-barrel-per-day pipeline are shown to range between $700,000 and $1.15 million per line-mile while the total costs for the 50,000-barrel-per-day pipeline are shown to range from $250,000 to $750,000 per line-mile.

Optimum combinations of pipeline size and number of pumping stations required for different line capacities over different distances and under different weather laying conditions were calculated with the aid of a computer program. The results of the calculations for oil pipelines under what can be called typical North Sea conditions, with the weather down-time factor of 3, are shown in

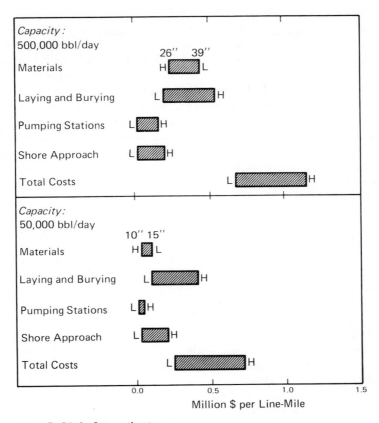

Source: Arthur D. Little, Inc., estimates.

Assume: $500/metric tons steel, laying and burying $200,000 per day, pumping station on production platform, shore approach $5 million.

[a]L: Pipeline 200 mi long, weather down-time factor is 1.
  H: Pipeline 25 mi long, weather down-time factor is 4.

Figure 3-20. Range of Typical Pipeline Costs[a] (in Millions of 1975 dollars).

Figure 3-18. It is assumed in these calculations that the maximum line size which can be laid is 42 inches even though the largest lines which have been laid to date are 36 inches. It is expected that laying of 42-inch line will be possible with the third-generation barges.

The importance of economies of scale in pipeline costs is clearly shown. A line with a capacity of 25,000 barrels over 100 miles would cost close to $1500 per daily peak barrel capacity, compared with $200 per daily peak barrel if the line could be sized to accommodate 500,000 barrels per day of oil.

To obtain a comparison between the alternatives of oil transportation by pipeline and by tanker, estimates of the costs for single-point mooring systems are shown in the same graph. The lower estimates shown are for a one-buoy system costing between $8 million and $10 million which, in the production range of 50,000 to 100,000 barrels per day, would require loading from the platform straight into a tanker moored to the buoy. For a range above 150,000 barrels per day, costs for an extended loading single-point mooring buoy system were used. This system, costing between $35 million and $40 million, has a storage capacity of 300,000 barrels, which allows the field to continue production over a number of days if the tanker cannot link up to the buoy because of adverse weather conditions. A complete comparison of the various types of transportation systems requires analysis of cash flows, which also allows for operating cost differentials. This analysis is performed in the analysis of minimum economic field sizes.

The results of similar analysis of the total costs for gas pipelines are shown in Figure 3–19. The same assumptions were made about materials and laying costs, shore approach costs, and the weather down-time factor. From the results, it is apparent that investment in the required pipeline will have to be about 25 percent higher if the same amount of British thermal units are to be transported over the same distance in the form of gas instead of in the form of oil.

Tanker Costs. The decision to link an offshore field with shore-based facilities by tanker, rather than through a pipeline, will be based upon considerations of field size (i.e., maximum number of years of production), production rate, cost of buffer storage, cost of pipeline construction, distance to shore, and onshore facilities at landfall. Since the OCS areas are rather large, substantial variations may occur in actual distances to be traveled, depending on the location of the well within the area and upon the distance to the nearest receiving terminal. The transportation from western and northern Alaskan fields, OCS areas 14 through 17, warrants special considerations: Are pipelines feasible for part of the trip, i.e., to southern Alaska, or should tankers be considered exclusively? As part of the data base for this study, the costs of oil transportation by tanker from offshore locations to continental United States ports have been estimated.

The transportation cost of crude oil by tanker can be expressed as a function of the distance traveled, the size of the tanker, and operating cost parameters of the site considered. These cost parameters vary with the flag of registration and with the year of construction and the year of operation of the tanker. Flag of registration for tankers operating between the United States outer continental shelf and United States ports is, of necessity, the United States flag (Jones Act). This implies that all cost parameters that depend upon the registration of the vessel, such as crew costs and insurance costs, have to be calculated for United States conditions. The year of ship construction determines the all-important yearly capital charges. Due to inflation in ship construction costs, older ships tend to be cheaper than new ships; hence, older ships show smaller capital charges than their younger sisters. We have assumed that new ships will be used; i.e., we have made a conservative, high estimate of tanker transporation costs.

The buildup of tanker costs is such that they are almost a linear function of distance traveled for any one ship size. Deviation from exact linearity is due to variations in payload for varying distances since longer distances require more bunkers to be carried, reducing cargo carrying capacity. Hence, deliverability is a function of distance traveled and the cost per ton of delivered oil increases slightly more than proportional with the length of the voyage. The other variable, ship size, gives rise to the well-known economies of scale in shipping, whereby costs per delivered ton decrease disproportionately with the size of the tanker up to a size of approximately 250,000 deadweight tons. Figure 3-21 is indicative of the relationship between cost and distance and cost and ship size, respectively.

For the purposes of this study, shipping costs are calculated for voyages of various lengths and for a selected number of vessel sizes below 150,000 dwt, the largest size that can be handled by a United States port and probably larger than the largest tanker that will be used on a shuttle service between offshore production and receiving terminals. With these calculated costs, graphs have been constructed (see Figure 3-22 and 3-23) that show delivered crude oil costs as a function of distance traveled one way, with the ship size as a parameter. Ship sizes not shown can be interpolated between those for which costs are given. These graphs enable the rapid determination of tanker transportation costs on any OCS route for any tanker size considered.

The transportation costs associated with each ship size-distance combination are considered to have two basic components: (1) fixed costs, and (2) voyage costs. The fixed costs of a tanker are independent of the trade she plies; the voyage costs depend upon both the tanker and the specific voyage considered

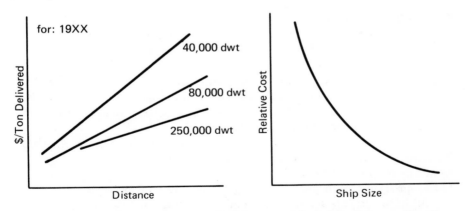

Figure 3-21. Shape of Functional Relationships Between Dollars per Ton Delivered and Distance and Relative Cost and "Ship Size"

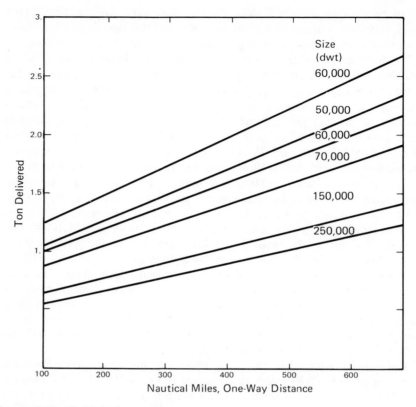

Source: Arthur D. Little, Inc., estimates.

Figure 3–22. Total Cost of Crude Oil Transportation in United States Flag Vessels in 1975 for Tanker Voyages of Less Than 1000 Nautical Miles One Way.

and have two basic components: bunker costs and port charges. Bunker consumption data used in this analysis assume steam turbine propulsion units, the dominant source of power for United States flag tankers. Bunker costs were estimated at $67.50 per long ton in 1975.

Port charges have been assumed nil at the loading ports, i.e., at the OCS locations; port charges at unloading ports have been taken as the average for United States ports. It should be noted that port charges on a yearly basis tend to become significant for the short voyages from most OCS locations.

All fixed costs are stated in terms of 1975 dollars, and the estimates made take into account current operating and financing practices in the United States. Operating practice is assumed at a level which might be encountered with a major oil company.

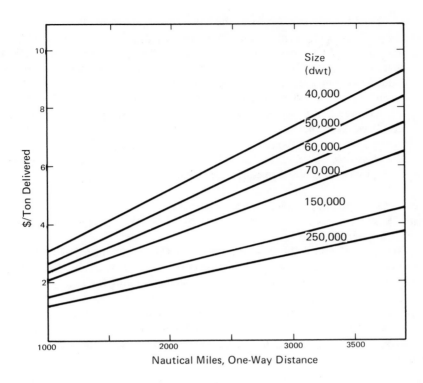

Source: Arthur D. Little, Inc., estimates.

Figure 3-23. Total Cost of Crude Oil Transportation in United States Flag Vessels in 1975 for Tanker Voyages up to 5000 Nautical Miles One Way.

The following fixed costs have been considered:

1. *Crew Costs.* These are as calculated by **MARAD** (Maritime Administration of the Department of Commerce) using a computer image program which takes into account details of a manning complement including overtime and fringe benefits. New wage contracts that went into effect on June 16, 1974 and June 16, 1975 have been taken into account.
2. *Insurance Costs.* These have been calculated from estimates of three components:
   a. Hull and machinery insurance—covers accidental damage to the vessel.
   b. Protection and indemnity insurance—covers such risks as injuries to crew or shore personnel, damage to third-party property, damage to cargo, oil pollution, etc. Most of this type of insurance is written through mutual ownership clubs, "P&I" clubs.

c. Total loss insurance—covers the risk of losing the vessel.

3. *Maintenance and Repair Costs.* These are calculated for the ship's mid-life year and assume a practical level of ship upkeep. Insurance costs reflect fleet operations with a reasonable history of claims and average levels of deductibles. The latter are included in the M&R costs.

4. *Capital Costs.* These have been based upon a review of published prices for delivered and on-order tankers from United States yards. The reported costs have been increased by 10 percent to allow for such costs as interest during construction, financing fees, legal and accountancy costs, etc.

5. *Miscellaneous Costs.* These include four principal items: shore overhead, stores, lubricating oils, and equipment rentals.

Some other operating parameters that have been taken into account are, for example, a vessel speed of 16 knots, port times of 1.5 days, vessel availability of 350 days per year, etc.

Figures 3–22 and 3–23 show the results of the calculations. The latter have been made with the aid of a computerized tanker cost model. Table 3–12 shows average distances and transportation costs in cents per long ton and in cents per barrel for selected voyages from offshore areas to onshore terminals. The transportation costs for LNG from Alaska (Point Gravina) to Los Angeles have been estimated at $1.00 per million Btu, in accordance with a recent study[3] done on the subject and assuming an internal rate of return of the LNG project of 15 percent.

Terminal Costs. For oil, the onshore terminal consists of a tank farm providing buffer storage capacity and a desulfurization and desalinization plant, if required. The investment costs on a per-barrel per-day basis for the desulfurization and desalinization plant are relatively small, around $6 to $10 per daily barrel capacity. These costs can be assumed to fall within the range of accuracy of any of the other major investment estimates for the development of offshore fields and are excluded explicitly from the analysis.

The tank farm, however, requires a significant capital outlay. For a typically-sized tank farm anywhere along the coast of the United States in the lower 48 states, the investment can be between $200 and $400 per peak daily throughput capacity. A tank farm constructed anywhere along the coast of Alaska will be more expensive because of higher costs for materials and construction, which will result in increases of between 10 and 20 percent of the costs for tanks, tank farm piping, miscellaneous equipment, and land development. The costs for the tank farm in a given location will depend on the number of days supply the tank farm is expected to accommodate, the type and size of tanks used, and the type of dikes used around the tanks. For instance, several states require steel dikes around the tanks instead of the cheaper earthen dikes, which can result in a significant cost increase for the total tank farm.

Table 3-12
Crude Oil Transportation Costs from OCS Areas to Likely Markets (1975 Dollars)
[Transportation costs in ¢/long tons[a] and distances in nautical miles (n.m.)]

To / From	New York (50,000 dwt)		Galveston (50,000 dwt)		Long Beach (150,000 dwt)		Seattle (150,000 dwt)	
	¢	n.m.	¢	n.m.	¢	n.m.	¢	n.m.
North Atlantic	150	300	495	2100	n.a.	n.a.	n.a.	n.a.
Middle Atlantic	140	250	445	1850	n.a.	n.a.	n.a.	n.a.
South Atlantic	220	600	335	1300	n.a.	n.a.	n.a.	n.a.
Gulf of Mexico	445	1850	150	300	n.a.	n.a.	n.a.	n.a.
California	1090	5100	990	4600	105	400	160	1000
Washington-Oregon	1180	5600	1090	5100	185	1000	120	500
Gulf of Alaska	n.a.	n.a.	n.a.	n.a.	300	2300	185	1250
Bristol Bay	n.a.	n.a.	n.a.	n.a.	310	2400	275	2050
Bering Sea	n.a.	n.a.	n.a.	n.a.	470	3900	360	2850
Chukchi Sea	n.a.	n.a.	n.a.	n.a.	490	4100	380	2850
Beaufort Sea	n.a.	n.a.	n.a.	n.a.	n.a.	n.a.	n.a.	3050

Source: Arthur D. Little, Inc., estimates.

[a]Conversion: 1 long ton ≅ 7.5 bbl.

Assuming a required capacity of 30 days of crude supply, a tank turnover factor of 5.7, an average tank size of 500,000 barrels per day, steel tanks with a floating roof, and every four tanks surrounded by an earthen dike results in estimates of $199 to $245 per barrel per day throughput capacity, depending on the overall size of the tank farm. Operating expenses range between 3.16¢ per daily barrel for the largest tank farm to 3.33¢ per barrel for the smallest tank farm.

The graphs in Figure 3–24 show investment and operating cost changes as functions of throughput.

Operating Costs

Operating costs will vary considerably for different platforms even if their capacities are the same. These costs depend on operating procedures and standards for the particular company, the reservoir characteristics, and the distance to the nearest supply base.

Based on an analysis of actual operating costs for the Gulf of Mexico, the Cook Inlet in Alaska, and the North Sea, the operating costs calculation is categorized into fixed costs, which can be assumed not to change for a given platform over its producing life, and variable costs, which will vary with the volume of oil and/or gas produced and the volume of water and/or gas injected into the reservoir. Fixed costs are divided into the following categories:

1. Wages and salaries
2. Provisions and catering
3. Transportation to and from the platform
4. Well workovers
5. Miscellaneous equipment maintenance
6. Insurance
7. Maintenance of electrical equipment
8. Overhead and contingencies

Wages and salaries, including payroll overhead, were found to be the same for the Gulf of Mexico as for the East and West Coast areas offshore the continental United States. Catering and quarters prove to be of the same order of magnitude per man-day as the cost of a first-class hotel in a major city.

Costs for transportation (by boat) of supplies and personnel will be similar all along the coast of the mainland, except for some areas where larger-than-average boats may be required. In those areas where frequent, and long, supply interruptions may be expected, the cost of larger-than-average supply vessels may be 20 to 25 percent above comparable costs in the Gulf of Mexico.

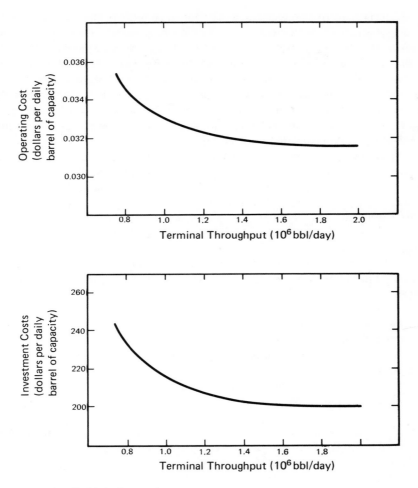

Source: Arthur D. Little, Inc., estimates.

Figure 3-24. Crude Oil Tank Farm Investment and Operating Costs as a Function of Throughput Capacity (1975 dollars).

Expected costs for supply and personnel transportation (by helicopter) in areas offshore Alaska are still higher. Helicopter charter time for work in Alaska is quoted as 3 to 5 times as high as the costs for comparative services in the Gulf of Mexico.

Well workover costs during the production period are a very significant factor in oil production economics, both onshore and offshore. They can range from several tens of thousands of dollars for small wells in the Gulf of Mexico to several hundreds of thousands of dollars for large wells in the rough environment of the North Sea. Wells located off the United States Atlantic and Pacific Coasts will show workover costs similar to those for wells in the Gulf of Mexico of comparable size. However, workover costs offshore Alaska are expected to be in the same range as workover costs in the North Sea, that is, at least 40 to 60 percent above those costs for wells in the Gulf of Mexico. Industry estimates the workover costs for subsea completions to be as much as 2.5 times those for conventional wells completed off a platform.

Miscellaneous equipment and its attendant maintenance are shown to be relatively constant percentages of the total cost of main platform equipment over a wide range of platform sizes. Yearly insurance is estimated at 2 percent of total capital investment in a platform. Costs of power plant maintenance can, quite consistently, be considered as a fixed amount per installed horsepower, which is directly related to the expected peak production capacity of either oil or gas.

Variable production costs comprise mainly energy and maintenance costs for oil pumps and gas compressors. The following variable costs are included in the analysis:

1. those directly related to oil production
2. those directly related to gas production
3. maintenance cost of gas injection equipment
4. gas injection operating costs
5. maintenance cost of water injection equipment
6. water injection proper

Directly related production costs for both oil and gas are proportional to daily produced volumes of oil and gas. Gas compressor maintenance costs are a function of the installed compressor capacity which, in turn, is determined by gas flow and required compression ratio.

The number of production personnel required on a platform is a function of the platform's peak capacity, ranging from 15 persons on a small platform to almost 100 on large platforms.

All cost elements are incorporated in a computerized operating cost model which allows for changing costs with changing well productivities and changing platform production capacities in various offshore areas.

Figure 3-25 is an illustration of fixed operating costs as a function of well productivity calculated for platforms with a productivity of 10,000 and 100,000 barrels per day in Alaska and in the Gulf of Mexico. For comparison purposes, the fixed operating costs of two 20,000-barrel-per-day production systems are included as well, one with subsea completion and the other with conventional platform completion.

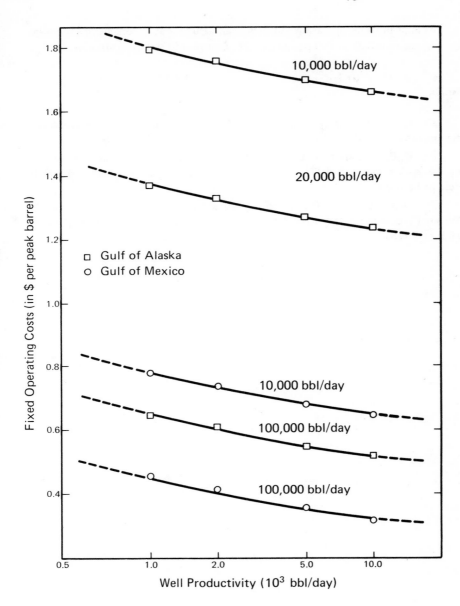

Source: Arthur D. Little, Inc., estimates.

Figure 3-25. Fixed Operating Costs as a Function of Platform Capacity (1975 dollars).

Annual pipeline operating cost has been assumed at 2 percent of the initial investment plus $20 to $30 per installed horsepower in pumping stations.

Notes

1.   Geological Survey Circular 725, "Geological Estimates of Undiscovered Recoverable Oil and Gas Resources in the United States," USGS 1975.

2.   P. B. Jenkins and A. L. Crockford, "Drilling Costs," paper presented at the spring meeting, 1975, of the Society of Petroleum Engineers of AIME held in London.

3.   Department of the Interior, "Alaskan Natural Gas Transportation Systems" Economic and Risk Analysis by The Aerospace Corporation, Draft of June 1975.

4

The Analyses

Analysis of Field Economics

Overview

The expected costs of exploration, development, and production of specific fields in the different OCS areas have been calculated based on the technical costs (presented in the section Cost Data in Chapter 3) as a function of the individual sizes and number of fields expected for these different OCS areas and by using the methodology described in Chapter 2. Investment costs and annual production costs by field size were determined in total and per daily unit produced. Costs were broken down into the following categories: exploration drilling costs, platform construction and installation costs, development well drilling costs, platform equipment costs, pipeline equipment costs, pipeline installation costs, gathering system costs, and onshore terminal costs.

To allow for cost differentials resulting from differences in weather and environmental conditions for the 17 different OCS areas, these areas were divided into eight groups (see Figure 4-1):

1. The Atlantic Coast area comprising BLM areas 1, 2, and 3
2. The Gulf of Mexico area, comprising the eastern, central, and western areas of the Gulf of Mexico
3. The southern Pacific Coast area, comprising southern California borderland and the Santa Barbara channel
4. The northern Pacific Coast area, comprising northern California and Washington-Oregon
5. The Gulf of Alaska
6. Lower Cook Inlet and Bristol Bay
7. Bering Strait and the Chukchi Sea
8. The Beaufort Sea

Minimum required price schedules were constructed based on the cash flows resulting from the cost calculations and allowing for royalty and tax payments in order to obtain an indication of what field sizes can be expected to be marginally economical under certain future cost/price conditions (see Figures 4-8 and 4-9).

77

Figure 4-1. Map of the Locations of the Outer Continental Shelf Areas.

Each calculation required the specification of basic parameter values for: (1) well productivity, (2) water depth, and (3) distance to shore. Average well productivity was assumed to be 500 barrels per day for oil and 20 million cubic feet per day for gas in the areas in the Gulf of Mexico and 2500 barrels per day for oil and 50 million cubic feet per day for gas in other OCS areas. The water depth and distance to shore representative of the different areas were taken from the Environmental Impact Statement[1] which shows potential drilling sites for each of the 17 areas.

To show how the minimum required price schedules can be expected to change with different values for the basic parameters, sensitivity tests were performed for the Gulf of Alaska, changing the values of the parameters over the ranges shown in Table 4-1. (See Figures 4-10 and 4-11 for the results of these sensitivity tests.)

The cost calculations for the base cases and for the sensitivity tests were done assuming field development from a fixed platform. To show how the costs can be expected to change with alternate types of field development for oil producing fields, minimum required price schedules were also developed for fields developed with subsea completions and a floating production station using a single-point mooring buoy and tankers to transport produced oil to shore.

Table 4-1
Minimum Required Price Calculations, Sensitivity Tests: Parametric Values

| | Water Depth (ft) | | | Distance to Shore (mi) | | | Well Productivity | | | | | | Required Rate of Return (%) | | |
| | | | | | | | Oil (bbl/day) | | | Gas (10⁶ CF/day) | | | | | |
	Low	Base	High	Low	Base	High	Low	Base	High	Low	Base	High	Low	Base	High
Gulf of Alaska	200	400	700	5	25	25	500	2500	10,000	20	50	100	10	15	25

Total Costs for Exploration and Development
for Individual Fields

As mentioned in the previous sections, an analysis of the costs required for exploration and development of individual oil and gas fields in the different OCS areas must allow for cost changes caused by differences in overall conditions. Within a particular area, the costs between individual fields can still be expected to vary significantly because of differences in:

1. The water depth at the location where the field has been found resulting in differences in platform construction, installation, and in pipe-laying costs.
2. The distance to shore, which will affect pipeline costs.
3. The depth and physical dimensions of the field, which will affect the development program (i.e., the number of platforms for a given amount of recoverable reserves); the amount of reserves that can be produced by one platform, depending on how deep the producing horizon is and on whether the producing formations are thinly spread out over a larger area or whether they are thick, which makes it possible to produce more of the reserves with the same platform.
4. The production characteristics of the reservoir itself which can affect the number of development wells that must be drilled, depending on the average well productivity and on the requirement for injection wells for water flooding.
5. The quality of the oil or gas, which can affect the amount of processing equipment required on the production platform, depending on how much stabilization and separation are required before the oil and/or gas can be transported to shore and whether associated gas has to be reinjected into the reservoir or has to be flared.

None of the values of these parameters, which all impact on the overall costs of field development, are known with certainty at the exploration stage. Therefore, a set of representative estimates is selected as base values for water depth, distance to shore, well productivity, reservoir depth, and barrels per acre or thousand cubic feet per acre for the different areas, as shown in Table 4–2.

Apart from the Gulf of Mexico and offshore California, where the average well productivity for oil and gas producing wells is approximately 500 barrels per day and 20 and 50 million cubic feet per day, respectively, base-case values have been selected to be 2500 barrels per day for oil and 50 million cubic feet per day for gas as the average for the other areas. A recovery of 50,000 barrels per acre was assumed for all areas except the Gulf of Mexico and the Atlantic, where 20,000 barrels per acre were used. For gas fields, a recovery of 100,000 MCF per acre was assumed for all areas.

Table 4-2
Base-Case Parameters

	Water Depth[a] (ft)	Distance to Shore (mi)	Well Productivity Oil (bbl/day)	Gas (10^3 CF/day)	Years Delay[b]	Bbl or CF per Acre Reserves 10^3 bbl/acre	10^6 CF/acre	Reservoir Depth (ft)
1. Atlantic Coast	400	75	2500	50	4	50	100	10,000
2. Gulf of Mexico	400	75	500	20	3	20	100	10,000
3. Pacific	600	15	500	50	4	20	100	10,000
4. Gulf of Alaska	400	25	2500	50	5	50	100	10,000
5. Lower Cook Inlet, Bristol Bay	200	15	2500	50	5	50	100	10,000
6. Bering Sea, Chukchi Sea	200	75	2500	50	5	50	100	10,000
7. Beaufort Sea	300	15	2500	50	5	50	100	10,000

Source: Arthur D. Little, Inc., estimates.

[a]This study considered only areas with a water depth not exceeding 600 feet (i.e., the strict definition of the outer continental shelf).

[b]Years delay from first discovery well until first production.

The values chosen for representative water depth and distances to shore for the different areas were based on information contained in the Environmental Impact Statement which also shows the location of the most likely drilling sites for the different areas (see Appendix B).

The depth and physical field dimensions, production characteristics, and quality of the oil and/or gas were assumed to be the same for all OCS areas. Consequently, calculations of total development and production costs were based on the assumption that similar fields—in terms of recoverable reserves, water depth, distance to shore, depth of producing horizon, and reservoir production characteristics—would require the same number of development wells and the same number of platforms with similar production equipment on those platforms in order to produce the fields. The only difference between otherwise similar fields, as shown in Table 4-2, is the different lead times between the first discovery well and the beginning of field production. These periods vary from 3 years for the Gulf of Mexico to 5 years for offshore Alaska. This assumption allows for differences in working conditions, a shorter working season in northern and polar areas, and longer distances from major supply centers.

In total exploration, development, and production costs, the different areas, starting with the most costly one, rank as follows:

1. Beaufort Sea and Chukchi Sea
2. Gulf of Alaska
3. Bering Sea and Bristol Bay
4. Lower Cook Inlet
5. Atlantic Coast
6. Gulf of Mexico
7. Pacific Coast

If water depth, shore distance, and average well productivity were the same for all areas, total costs in the most expensive area, the Beaufort Sea, would be approximately 5 times as great as in the least expensive area, the Pacific.

When expected differences in water depth and distance to shore between the areas are considered, the expected costs become those shown in Figures 4-2 and 4-3 for typical 150-million-barrel and 2½-trillion-cubic-foot oil and gas fields, respectively. The rank ordering in terms of the costliness mentioned above changes by using different water depths and distances-to-shore for the different areas. The development of the "typical" field offshore California, in the case of oil, at a total cost of $156 million, is shown to be more expensive than that in the Gulf of Mexico, the Atlantic Coast, and the Lower Cook Inlet with total costs of $152 million, $135 million, and $142 million, respectively. This is mainly because of much higher development drilling costs and higher platform costs offshore California as a result respectively of the larger number of wells which are required, given the lower well productivity, and because of deeper water.

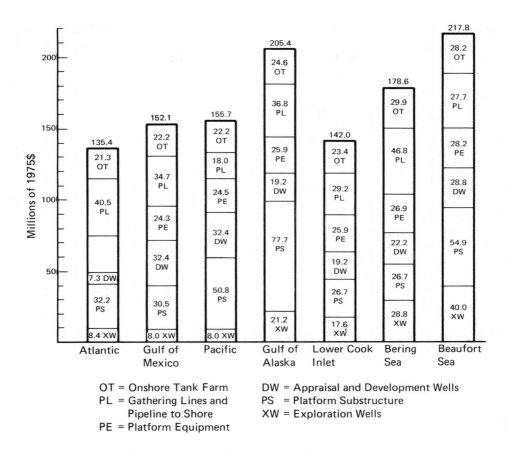

Figure 4-2. Exploration Drilling Costs and Field Development Costs for a "Typical" Field with 150 Million Barrels Recoverable Reserves—Oil

Note: See Table 4-2 for base-case assumptions.

Source: Arthur D. Little, Inc., estimates.

In the case of gas (see Figure 4-3), the Atlantic and the Gulf of Mexico are shown to be considerably more expensive than the Lower Cook Inlet, $179.7 million and $165.0 million versus $136.0 million, respectively. The Gulf of Alaska and the Bering Sea are shown as more expensive than the Beaufort Sea, $217.3 million and $215.9 million versus $211.8 million, respectively. In the case of the Atlantic Coast and the Gulf of Mexico, the higher expenditures are caused mainly by the higher pipeline costs since it is expected that fields in the Atlantic and in the Gulf of Mexico will require pipelines to shore averaging 75

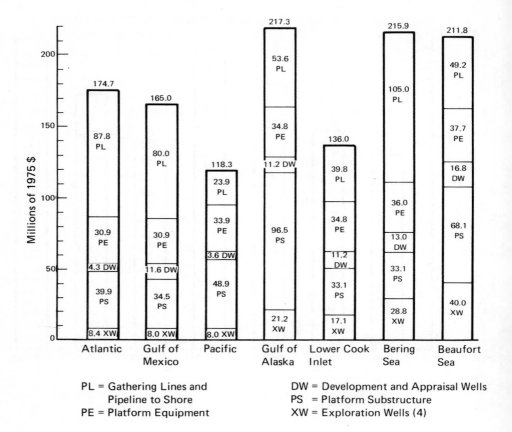

Millions of 1975 $

PL = Gathering Lines and
Pipeline to Shore
PE = Platform Equipment

DW = Development and Appraisal Wells
PS = Platform Substructure
XW = Exploration Wells (4)

Source: Arthur D. Little, Inc., estimates

Note: See Table 4-2 for base-case assumptions.

Figure 4-3. Exploration Drilling Costs and Field Development Costs for a "Typical" Field with 2500 Billion SCF Recoverable Reserves— Gas.

miles compared to averages of 15 miles expected for the Lower Cook Inlet area. Longer pipelines to shore for the Bering Sea also result in higher costs for that area compared to the Beaufort Sea. The higher total field development costs for the Gulf of Alaska, compared with the Beaufort Sea, result from the expected deeper water of the Gulf of Alaska (400 feet versus 300 feet, respectively), which results in significantly higher total platform cost estimates.

The costs shown in Figures 4-2 and 4-3 must be regarded as the *minimum investment costs* required. Onshore terminals for oil include cost estimates for a tank farm but exclude potentially required processing facilities such as desalinization or desulfurization plants which depend on the quality of the crude. In the case of gas, onshore natural liquid gas processing facilities have also been excluded. The exploration costs, which are included in the total costs, include four exploration wells which can be considered to be the minimum exploration costs that would be allocated to a field.

Figures 4-4 and 4-5 show the relative contribution of the different cost categories to total exploration drilling and development costs calculated for the typical oil and gas fields for each major OCS area. It is clear from these two figures that the relative exploration drilling costs increase as one moves into more remote and more hostile areas. The costs of the typical four exploration wells are estimated to range from 5 to 7 percent of total costs in the areas off the Atlantic and Pacific Coasts and in the Gulf of Mexico, while their relative costs are estimated to range from 10 to 20 percent in the offshore areas of Alaska. In other words, not only are general cost levels higher in those remote and more hostile areas, but that portion of the total capital required for exploration and field development which has to be *risked* in exploration drilling is also higher. Companies can be expected to have to invest 3½ times more capital in field exploration and development in the Gulf of Alaska than in the Pacific, and they can be expected to have to risk at least 4 to 5 times as much capital in the exploration drilling phase. In addition, it will take about twice as long before they realize their first production.

Platform costs are extremely sensitive to water depth when comparing the Gulf of Mexico costs for 400 feet with costs for the Pacific, where an average water depth of 600 feet is expected. The platform costs as a portion of total investment range from 15 percent for an area with shallow waters and a relatively large investment in the pipeline to shore to 40 percent for a high-cost area, such as the Gulf of Alaska with deep water (400 feet) and with relative closeness to the Coast (25 miles).

Development drilling costs in the cases shown ranged from 5 to 20 percent of total cost. These costs are sensitive to the average well productivity, as illustrated by the difference between the development drilling costs for the Gulf of Mexico and those for the Atlantic offshore areas, where they were estimated at $31.4 million and $7.3 million, respectively, for well productivities of 500 and 2500 barrels per day of oil.

Production equipment costs range from 12 to 20 percent of total costs in the case of oil and from 15 to 30 percent of total costs in the case of gas. For oil, costs for the pipeline to shore and the onshore tank farm range from 30 percent for the Pacific Coast and the Gulf of Alaska, which are close to shore, to 45 percent in the case of the Bering Sea and the Atlantic, where it is expected that pipelines will be at least 75 miles long. For gas, construction cost

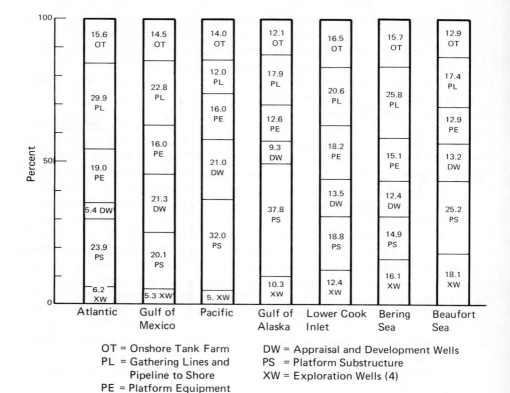

OT = Onshore Tank Farm   DW = Appraisal and Development Wells
PL = Gathering Lines and   PS = Platform Substructure
       Pipeline to Shore   XW = Exploration Wells (4)
PE = Platform Equipment

Source: Arthur D. Little, Inc., estimates.

Note: See Table 4-2 for base-case assumptions.

Figure 4-4. Percentage Distribution of Exploration Drilling Costs and Field Development Costs for a "Typical" Field with 150 Million Barrels of Recoverable Reserves—Oil.

of the pipeline to shore requires by far the largest investment, about 50 percent of total costs.

Unit Costs of Production: Economies of Scale

The costs for different fields in the same area and for the same fields between areas must also be compared in terms of dollars per unit of maximum field production capacity. Figures 4-6 and 4-7 show the unit costs by category for:

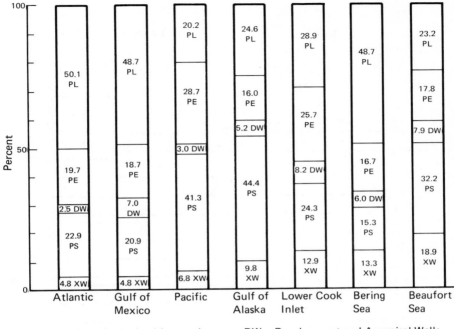

PL = Gathering Lines and
     Pipeline to Shore
PE = Platform Equipment

DW = Development and Appraisal Wells
PS = Platform substructure
XW = Exploration Wells (4)

Source: Arthur D. Little, Inc., estimates.

Note: See Table 4-2 for base-case assumptions.

Figure 4-5. Percentage Distribution of Exploration Drilling Costs and Field Development Costs for a "Typical" Field with 2500 Billion SCF of Recoverable Reserves—Gas.

1. the unit costs to drill four exploratory wells
2. the total of platform construction and installation costs, the production equipment costs, and development drilling costs
3. the total costs for the pipeline to shore and, in the case of oil, the costs for an onshore tank farm

These costs are shown for oil fields of 45 million barrels, 150 million barrels, and 2 billion barrels of recoverable reserves (Figure 4-6), for gas fields of 0.25 trillion cubic feet, 1 trillion cubic feet, and 10 trillion cubic feet of recoverable reserves (Figure 4-7), and for the average expected well productivity in each area.

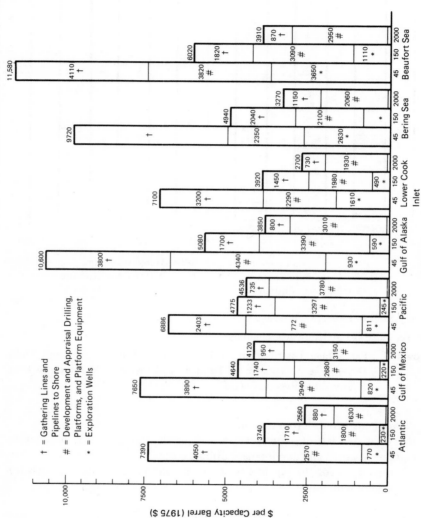

Source: Arthur D. Little, Inc., estimates.

Note: See Table 4–2 for base-case assumptions.

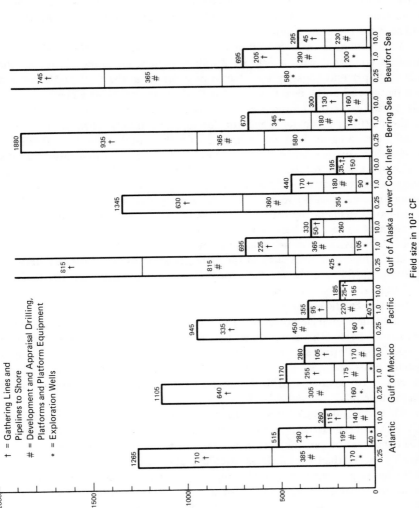

Field size in 10^{12} CF

Source: Arthur D. Little, Inc., estimates.

Note: See Table 4-2 for base-case assumptions.

Figure 4-7. Exploration and Field Development Costs per SCF Production Capacity for Three Different Field Sizes—Gas.

Again, as in the case of the total cost estimates, it must be emphasized that the costs have been calculated for *typical* fields using expected values for water depth, distance to shore, and average well productivity for the particular areas. If the expected values were the same for all areas, then the unit cost for the Beaufort Sea would be about 5 times higher than the unit cost shown for the Pacific Coast. The geographical conditions assumed for the differently-sized fields in the same area were identical, and the unit costs shown for these different field sizes are therefore comparable.

The economies of scale within the same area, when a field of 45 million barrels is compared with a field of 2 billion barrels, are shown to reduce the total costs per unit capacity by a factor of 2 for oil in the Gulf of Mexico and by a factor of 3 in the Beaufort Sea. The Gulf of Mexico shows a higher cost per unit capacity for the required construction and installation of the platforms, for the platform equipment, and for development drilling when the unit costs of the 2-billion-barrel field are compared with the unit costs of the 150-million-barrel field. This is explained by the longer lead time required to drill all the development wells for the 2-billion-barrel field (given the assumption of an average well productivity of 500 barrels per day) which results in a longer producing life and a smaller required overall capacity than are necessary for the 2-billion-barrel fields in the other OCS areas, where the average well productivity was assumed to be 2500 barrels per day and where, as a consequence, the smaller number of wells allows more rapid field development.

In the high-cost areas such as the Bering Sea and the Beaufort Sea, unitized exploratory drilling costs for the expected four exploratory wells per field are more than the total unit cost to explore for and develop a large field off the Atlantic and Pacific coasts.

Economies of scale are shown to be much more pronounced for gas fields than for oil fields. This is explained by the relatively larger investment required in the transportation system to shore for gas fields. In the Beaufort Sea, unit capacity costs are more than 7 times higher for the smallest fields of 0.25 trillion cubic feet as compared to the largest field with recoverable reserves of 10 trillion cubic feet.

When the unit capacity investments required for oil and gas are compared on a BTU-equivalent basis, assuming that 1 barrel of crude oil is equivalent to 6000 cubic feet of gas, the unit investment costs for smaller gas fields are considerably higher than for oil fields of comparable size, while unit capacity costs for the larger gas fields are considerably smaller than the unit capacity costs for the larger oil fields. In comparing estimates for the 0.25-trillion-cubic-foot field for gas with the 45-million-barrel field for oil in the Beaufort Sea area, the gas field will require approximately $13,000 per barrel capacity, in terms of crude oil equivalent,[a] as compared with almost $12,000 per barrel capacity for the

---

[a] 6000 cubic feet of natural gas is roughly 1 barrel of crude oil equivalent on a thermal basis.

oil-producing field. However, in comparing the unit costs of a giant gas field of 10 trillion cubic feet of recoverable gas with the unit costs of a giant oil field of 2 billion barrels of recoverable oil, the investment costs per barrel of crude oil equivalent for the gas field are about $1800 compared with the $4000 per capacity barrel for the oil field.

Minimum Required Price

Given estimates for technical costs for each of the 17 different OCS areas and estimates for the time required to bring a discovered field into production, one can calculate the minimum price that a company must require in order to cover the costs for exploration, development, and production of oil and gas fields of different sizes and under different conditions.

As explained in Chapter 2, the minimum required price resulted from a discounted cash flow calculation, allowing for royalty and tax payments over the producing life of the fields. The price that is calculated can be considered to be the break-even price which allows companies to cover a nominal portion of the exploration costs plus the development and production costs while making a required return on their capital.

This rate of return on capital will vary depending on how a particular company assesses the riskiness of the particular area where it is trying to acquire the right to explore for oil and gas fields. It is beyond the scope of this book to try to show what this rate of return is or will be. Therefore, we have chosen to develop the minimum required prices for different rates of return ranging from 10 percent per annum to 25 percent per annum.

Figures 4-8 and 4-9 show the minimum required prices for oil and gas fields, respectively. As in the case with the per unit capacity costs, the minimum required prices are shown for three different field sizes—45 million barrels, 150 million barrels, and 2 billion barrels in the case of oil, and 0.25, 1, and 10 trillion cubic feet in the case of gas. The shorter lead times between the first discovery and first production, as assumed for the Gulf of Mexico when compared with the Atlantic and the Pacific coast areas, are shown to result in a lower required price for the Gulf of Mexico for the cases of both oil and gas.

The differences between minimum required prices calculated for the same field with different rates of return are shown to be quite significant. A 45-million-barrel field in the Lower Cook Inlet area would be economical with a minimum required price of $10.63 if the company were satisfied with a 10 percent rate of return. When a 15 percent rate of return is required, however, the field could not be developed at a required price of $14.26 if the regulated price were around the present level of $11.28. At a required rate of return of 25 percent, even a field of 150 million barrels might be considered uneconomical given a minimum required price of $11.99.

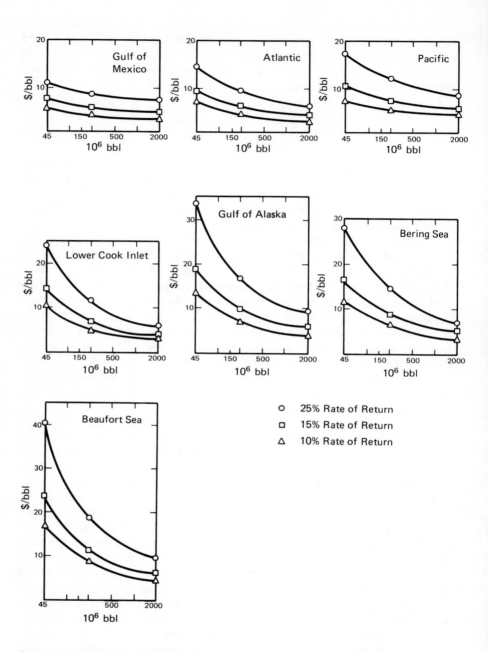

Source: Arthur D. Little, Inc., estimates.

Figure 4-8. Minimum Required Prices as a Function of Field Size, Calculated with Different Rates of Return—Oil.

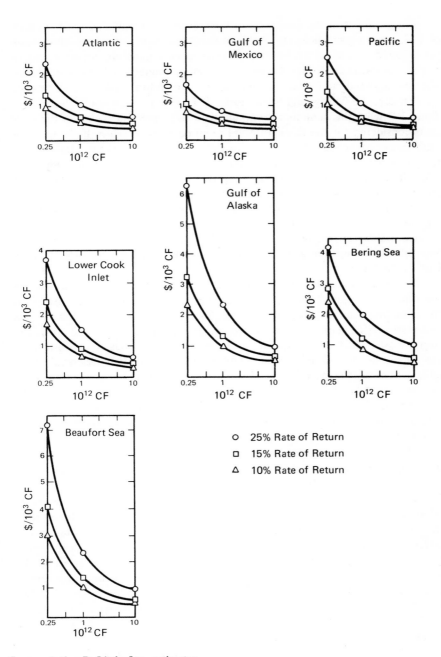

Source: Arthur D. Little, Inc., estimates.

Figure 4-9. Minimum Required Prices as a Function of Field Size with Different Rates of Return—Gas. (1975 $)

For gas fields in the Lower Cook Inlet, assuming a regulated wellhead price of 52¢, even a giant field of 10 trillion cubic feet would not be economical if the required rate of return were 25 percent, given a minimum required price of 63¢ per thousand cubic feet. However, with minimum required prices of 33¢ and 41¢ per thousand cubic feet and required rates of return of 10 and 15 percent, respectively, the field would be economical. To illustrate the results, we show in Table 4-3 what the minimum economic field size would be in each of the seven areas considered for the three different required rates of return if the wellhead price is 75¢ per thousand cubic feet for gas and $12 per barrel for oil. As shown, for gas, assuming a required rate of return of 15 percent, only fields larger than 500 billion cubic feet of recoverable reserves would be developed in the Gulf of Mexico, a low-cost area, and only fields larger than 4000 billion cubic feet would be developed in the Gulf of Alaska, a high-cost area. For oil, again assuming a required rate of return of 15 percent, only fields larger than 17 million barrels of recoverable reserves would be economical to develop in the Gulf of Mexico and only fields with more than 97 million barrels of recoverable reserves in the Gulf of Alaska.

The minimum economic field size in the different areas is very sensitive to the value of the required rate of return used in the calculations. In the case of gas, the minimum economic field size will be about 8 times larger, 2500 billion cubic feet versus 310 billion cubic feet of recoverable reserves, if required rates of return are used which differ by a factor of 2.5, that is, 25 versus 10 percent. In the case of oil, in the Gulf of Mexico the minimum economic field size would be 11 million barrels of recoverable reserves, assuming a required rate of return of 10 percent, and 47 million barrels, assuming a required rate of return of 25 percent; i.e., the minimum economic field size is larger by a factor of 4 if the required rate of return used to calculate this minimum economic field size is larger by a factor of 2.5.

The results of sensitivity tests, assuming different values of the different parameters and different development programs for the Gulf of Alaska area, are shown in Figures 4-10 and 4-11.

The sensitivity to changes in assumed well productivity is shown to be relatively small, especially for the smaller fields. For a 45-million-barrel field, the minimum required price for the Gulf of Alaska, given the assumptions on water depth and distance to shore and number of years delay until first production, would be $19.44 per barrel if a well productivity of 10,000 barrels per day were assumed and $22.99 per barrel if a 500-barrel-per-day well productivity were assumed. The minimum required price for well productivities of 10,000 barrels per day and 500 barrels per day for a 2-billion-barrel field would be $4.79 and $9.90, respectively; in other words, a well productivity of about 20 times as high would reduce the minimum required price only by a factor of 2.

The effect of assumed changes in cost/price relationships is shown to be quite significant. Assuming prices to increase relative to costs at 5 percent a

Table 4-3
Minimum Economic Field Size[a]

| | Rate of Return | | | | | | Assumptions | | | | | |
| | Gas (billions of cu ft) Wellhead Price $.75/10³ CF | | | Oil (millions of bbl) Wellhead Price $12/bbl | | | Average Well Production | | Distance to Shore | Water Depth | Years Delay[b] | Recovery |
	10%	15%	25%	10%	15%	25%	(bbl/day)	(10³ CF/day)				(10³ bbl/acre)	10⁶ CF/acre
Atlantic	180	290	660	17	26	70	2500	50	75	400	4	50	100
Gulf of Mexico	120	185	400	11	17	47	500	20	75	400	3	20	100
Pacific	220	300	770	19	37	125	500	50	15	600	4	20	100
Gulf of Alaska	660	1100	5400	60	97	425	2500	50	25	400	5	50	100
Lower Cook Inlet	370	560	1550	37	58	150	2500	50	15	200	5	50	100
Bering Sea	600	930	4400	49	80	260	2500	50	75	200	5	50	100
Beaufort Sea	850	1600	6400	80	135	560	2500	50	15	300	5	50	100

Source: Arthur D. Little, Inc., estimates.

[a]In recoverable reserves.

[b]Number of years between first discovery well and first field production.

[c]Field recovery in thousand barrels per acre or million cubic feet per acre.

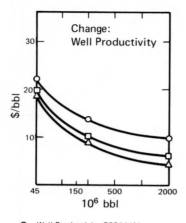

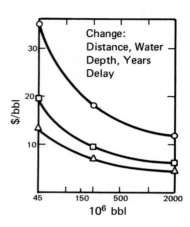

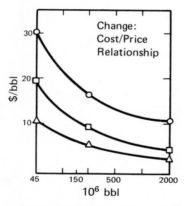

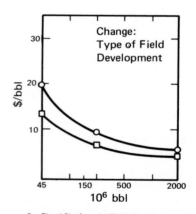

Source: Arthur D. Little, Inc., estimates.

Figure 4-10. Gulf of Alaska, Oil—Results of Sensitivity Tests on Minimum Required Prices (1975 $).

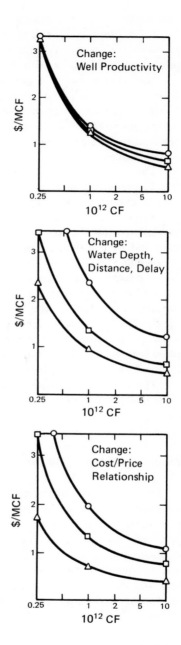

o Well Productivity 20 10^6 CF/day

□ Well Productivity 50 10^6 CF/day

△ Well Productivity 100 10^6 CF/day

		Water Depth (ft)	Distance to Shore (mi)	Delay (yr)
o	High	700	50	6
□	Base Case	400	25	5
△	Low	200	5	4

o Costs Increase Relative to Price at 5% per Year

□ Costs and Price Change at the Same Rate

△ Price Increases Relative to Costs at 5% per Year

Source: Arthur D. Little, Inc., estimates.

Figure 4-11. Results of Sensitivity Tests on Minimum Required Prices Using a Required Rate of Return of 15 Percent per Year.

year would reduce the minimum required price from $5.67 to $3.75 per barrel for a 2-billion-barrel field. Assuming costs to increase at 5 percent per year relative to prices over the life of the field would increase the minimum required price from $5.67 per barrel to $10.70 per barrel for the 2-billion-barrel field.

Assuming a greater water depth, a larger distance to shore, and a longer delay time between first discovery and first production also has a significant effect on the minimum required prices. The effect of the different rates of return has already been discussed. Assuming different field development pro-grams, using subsea completions and a floating platform, has a significant effect on the smaller field sizes.

The sensitivity tests for gas fields show the same results as obtained for oil (see Figure 4-11). Increasing the average well productivity by 5 times, from 20 million cubic feet per day to 100 million cubic feet per day, decreases the minimum required price by only 30 percent, from $0.56 per thousand cubic feet to $0.70 per MCF in the case of the largest field size assumed (10 trillion cubic feet) and by less than 1 percent in the case of the smallest field assumed (250 billion cubic feet).

If prices increase relative to costs at 5 percent per year, then the minimum required price will decrease from $0.64 per thousand cubic feet to $0.37 per thousand cubic feet for the 10 trillion cubic feet field and from $3.45 to $1.74 for the 250 billion cubic feet field. If costs increase relative to prices at a rate of 5 percent per year, then the minimum required price will be $1.00 per thousand cubic feet for the 10-trillion-cubic-foot field and $4.37 per thousand cubic feet for the 250-billion-cubic-foot field.

The sensitivity to changes in assumed values for water depth, distance to shore and delay until first production is shown to be very significant. The mini-mum required price almost doubles between the base case and high case.

Projections of Future Oil and Gas Production

As described as part of the methodology (Chapter 2), two types of information were used to simulate the exploration and subsequent development and produc-tion activities for different areas on the OCS for which lease sales have been proposed through 1978: (1) probabilistic information on resource base size, structure size distribution, and the distribution of possible fills of structures with oil or gas; and (2) the information developed on the economics of explora-tion and development activities. It should be emphasized that the projected production streams are functions of the proposed lease sale schedule, which is shown in Table 4-4. It can be expected that if oil or gas is found in any of the areas considered in the analysis, the first lease sale will be followed by a second and maybe a third lease sale in the period covered by the analysis, i.e., 1975 to 1990. Therefore, it can be expected that for those areas where the possibility

Table 4–4
Lease Sale Schedule and Millions of Acres Leased as Assumed for
Area Simulations

Area Name	Time of Lease Sale and Millions of Acres Leased				
	1974	1975	1976	1977	1978
1. North Atlantic			0.4		
2. Mid-Atlantic			0.4		
3. South Atlantic			0.4		
4. Eastern Gulf	0.5				
5. Central and Western Gulf	1.2	1.2			
6. South California and Santa Barbara Channel				0.6	
7. Northern California and Washington-Oregon					
8. Gulf of Alaska, East			0.4		
9. Gulf of Alaska, Kodiak				0.4	
10. Gulf of Alaska, Aleutian Shelf					0.3
11. Lower Cook Inlet				0.5	
12. Outer Bristol Basin					0.5
13. Bering Sea, Norton Basin				0.5	
14. Bering Sea, St. George				0.5	
15. Chukchi Sea, Hope Basin					0.5
16. Beaufort Sea				0.5	

Source: Arthur D. Little, Inc., estimates.

of substantial production levels is shown, these same production levels will not decline between 1985 and 1990 as shown in the projections, but most probably will stay level or even increase.

Base Projections

Results of an area projection for the Gulf of Alaska are shown in Tables 4–5 and 4-6. The possible production of oil and gas and the possible annual expenditures required to find, develop, and produce that oil and gas and to transport it to the nearest point of sale have been calculated for 10 benchmark years for different price categories and at different levels of confidence within each price category.

Table 4–5

Projections of Oil Production Levels under Different Price Scenarios and at Different Levels of Confidence as Resulting from Lease Sales through 1978 — Gulf of Alaska, East

			Expected Price $4.50/bbl or $.75/10³ CF (No Production)			
			Expected Price $7.50/bbl or $1.25/10³ CF			
			Oil Production (10⁶ bbl/yr)			
Confidence Level	1980	1981	1982	1983	1985	1990
5%	.00	127.17	194.57	265.42	359.28	258.02
25%	.00	.00	.00	60.07	80.88	60.38
50%	.00	.00	.00	.00	.00	.00
75%	.00	.00	.00	.00	.00	.00
95%	.00	.00	.00	.00	.00	.00
			Expected Price $12/bbl or $2/10³ CF			
5%	.00	127.17	194.57	265.42	381.72	284.56
25%	.00	.00	52.53	76.46	102.80	71.56
50%	.00	.00	.00	18.48	35.34	26.34
75%	.00	.00	.00	.00	.00	.00
95%	.00	.00	.00	.00	.00	.00
			Expected Price $18/bbl or $3/10³ CF			
5%	.00	127.17	194.57	265.42	391.72	284.56
25%	.00	.00	52.53	76.46	102.80	73.54
50%	.00	.00	.00	18.86	35.39	26.67
75%	.00	.00	.00	.00	.00	.00
95%	.00	.00	.00	.00	.00	.00
			Expected Price $24/bbl or $4/10³ CF			
5%	.00	127.17	194.57	265.42	391.72	284.56
25%	.00	.00	52.53	78.56	102.80	73.54
50%	.00	.00	.00	22.39	35.78	27.17
75%	.00	.00	.00	.00	.00	.00
95%	.00	.00	.00	.00	.00	.00

Source: Arthur D. Little, Inc., estimates.

As shown in Tables 4-5 and 4-6, assuming a price of $4.50 per barrel for oil and $.75 per thousand cubic feet for gas landed in California, no oil or gas will be developed even if there is some in the eastern part of the Gulf of Alaska. If the expected price for oil and gas landed in California is $7.50 per barrel and $1.25 per thousand cubic feet, respectively, then, at a confidence level of 50 percent, one still cannot expect any oil or gas production to occur. Only at the lower confidence levels of 25 percent can one expect some oil production to result from an exploration and development effort in that area. At the 25 percent confidence level, this production will reach its peak of at least 8 million barrels per year in 1985; at a 95 percent confidence level it will reach at least 360 barrels per year in

Table 4-6
Projections of Gas Production Levels under Different Price Scenarios and at Different Levels of Confidence as Resulting from Lease Sales through 1978 — Gulf of Alaska, East

Expected Price $4.50/bbl or $.75/10³ CF
(No Production)

Expected Price $7.50/bbl or $1.25/10³ CF
(No Production)

Expected Price $12/bbl or $2/10³ CF

			Gas Production (10⁶ CF/yr)			
Confidence Level	*1980*	*1981*	*1982*	*1983*	*1985*	*1990*
5%	.00	209.01	313.76	508.64	522.42	522.42
25%	.00	.00	.00	.00	.00	.00
50%	.00	.00	.00	.00	.00	.00
75%	.00	.00	.00	.00	.00	.00
95%	.00	.00	.00	.00	.00	.00

Expected Price $18/bbl or $3/10³ CF

5%	.00	209.01	337.80	508.64	633.80	633.80
25%	.00	.00	.00	102.12	121.74	131.74
50%	.00	.00	.00	.00	38.79	38.79
75%	.00	.00	.00	.00	.00	.00
95%	.00	.00	.00	.00	.00	.00

Expected Price $24/bbl or $4/10³ CF

5%	.00	209.01	337.80	508.64	646.39	646.39
25%	.00	.00	.00	102.12	154.30	154.30
50%	.00	.00	.00	.00	47.93	47.93
75%	.00	.00	.00	.00	.00	.00
95%	.00	.00	.00	.00	.00	.00

Source: Arthur D. Little, Inc., estimates.

1985. As shown in Table 4-6, at a landed price in California of $1.25 per thousand cubic feet, one cannot expect any gas to be developed and produced if found in the Gulf of Alaska. Only if the expected price for gas landed in California is close to $2 per thousand cubic feet can one expect production on the Gulf of Alaska areas to accrue at a 5 percent confidence level at a peak of at most 522 billion cubic feet per year in 1985. At an expected price of $3 per thousand cubic feet, production levels in the eastern part of the Gulf of Alaska would accrue in 1985 at levels of at most 739 billion cubic feet per year at a confidence level of 50 percent; at most 132 billion cubic feet per year at a confidence level of 25 percent, and 634 billion cubic feet per year at a confidence level of 5 percent.

The prices shown in Tables 4-5 and 4-6 are minimum required prices to cover the nominal amount of exploration drilling costs, i.e., the cost for drilling four exploratory wells, all development and production costs, plus the charges for the transportation of the crude oil and gas to the closest market, which has been assumed to be California. Lease bonuses are excluded. For the same level of confidence, more production can be expected if the expected prices are higher; more previously marginal fields are developed and produced as the price increases. It should be emphasized that the incremental amount of production shown to result from a higher expected price level cannot be interpreted as showing all the price sensitivity of production in a particular area. It shows only the incremental production which can be expected to result from a successful exploration effort during *the first exploration period* (after the first lease sale) if the industry is confident that future prices will be at the higher level instead of at the lower level.

The lease sale process is assumed to be efficient in selecting the bigger structures which have a higher chance of containing a large oil or gas field and leaving out most of the smaller structures which would only be economical at the higher price levels shown in Figure 4-11. These smaller structures will become the subject of companies' exploration efforts in subsequent lease sales. Having obtained a better understanding of the areas' petroleum geology and having invested in the areas' general infrastructure, companies will have more reason to explore these smaller structures if they expect to find commercial fields. The higher the future expected price levels, the smaller the structures which companies will be interested in exploring and developing. As discussed in the section of this chapter "Supply Curves for New Offshore Areas" the supply curves for new OCS areas, showing the total sensitivity of future production levels to companies' price expectations, can be derived only from an analysis based on this full cycle of lease sales.

The results of the probabilistic production projections for the 16 different lease sales assumed to be held through 1978 (see Table 4-4) combine to obtain production projections at different confidence levels for the different areas off the coast of the United States and for the benchmark years 1980, 1985, and 1990. For each benchmark year, the probabilistic forecasts are combined to obtain the joint probability distributions of total possible production levels for consolidated areas of the Atlantic Coast, the Gulf of Mexico, the Pacific Coast, or offshore Alaska. Figures 4-12 and 4-13 show the results when probabilistic projections are combined for oil and gas production for the different combined areas.

The expected price levels assumed for these aggregated probabilistic forecasts were $12 per barrel for oil for all areas; $1.25 per thousand cubic feet for gas off the Atlantic Coast, in the Gulf of Mexico, and off the Pacific Coast; and $2 per thousand cubic feet for gas produced and transported to California from Alaskan offshore areas. In other words, Figures 4-12 through 4-15 show for combined areas the expected total production levels resulting, at different levels of confidence, from lease sales through 1978 when wellhead prices for oil range between $11 and $12 per barrel and when the wellhead price for gas is

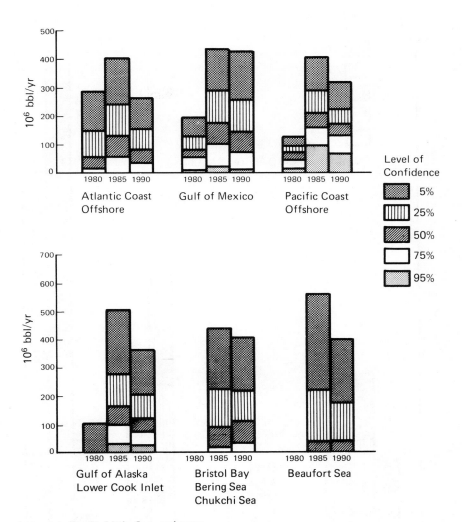

Source: Arthur D. Little, Inc., estimates.

Figure 4-12. Estimates of Possible Annual Production Levels at Different Confidence Levels from Areas Leased or to Be Leased Through 1978 on the Outer Continental Shelf of the United States—Oil.

approximately $1.25 per thousand cubic feet. It is apparent from these figures that the full range of all possible outcomes from an accelerated lease sale through 1978 in terms of potential oil production in 1985 reaches from a total of at most 570 million barrels per year (1.6 million barrels per day) at a high level of

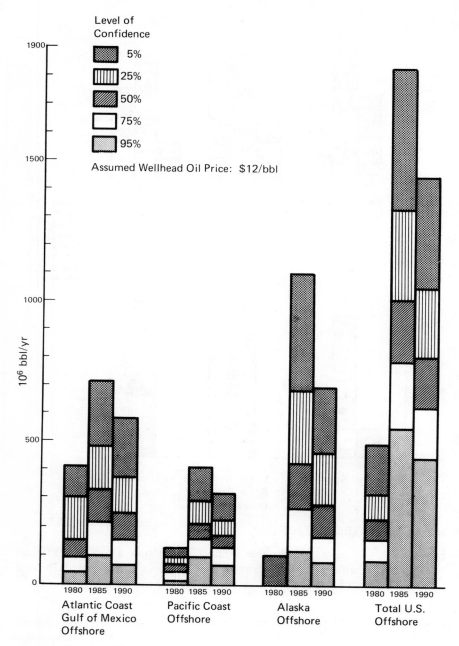

Source: Arthur D. Little, Inc., estimates.

Figure 4-13. Estimates of Possible Annual Production Levels at Different Confidence Levels from Areas Leased or to Be Leased Through 1978 on the Outer Continental Shelf of the United States—Oil.

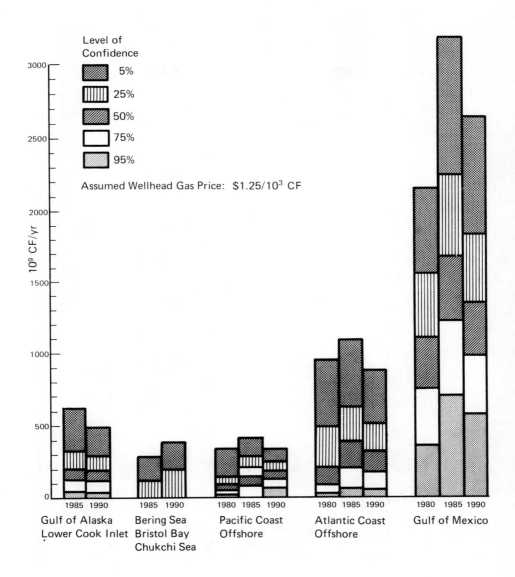

Figure 4-14. Estimates of Possible Annual Production Levels at Different Confidence Levels from Areas Leased or to Be Leased Through 1978 on the Outer Continental Shelf of the United States—Gas.

Note: At the assumed wellhead price of $1.25/10³CF, no gas will be produced from Beaufort Sea.

Source: Arthur D. Little, Inc., estimates.

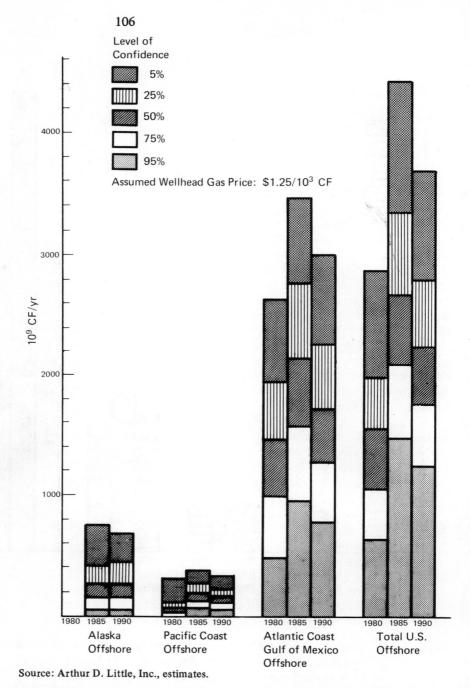

Source: Arthur D. Little, Inc., estimates.

Figure 4-15. Estimates of Possible Annual Production Levels at Different Confidence Levels from Areas Leased or to Be Leased through 1978 on the Outer Continental Shelf of the United States—Gas.

confidence (95 percent) to a high of at most 1840 million barrels per year (5.1 million barrels per day) at low levels of confidence (5 percent) if wellhead prices are expected to be approximately $12 per barrel. Potential production of gas for all areas combined in 1985 ranges from at most 1500 billion cubic feet per year (4.11 billion cubic feet per day) at a high level of confidence (95 percent) to a high of at most 4500 billion cubic per year (12.3 billion cubic feet per day) at a low level of confidence (5 percent).

Expected Production under Alternate Scenarios

The expected values of the projected production level for oil and gas resulting from projections for the individual areas for a range of different price levels, $4.50 per barrel to $18 per barrel for oil and $.75 per thousand cubic feet to $3 per thousand cubic feet for gas, are shown in Tables 4-7 and 4-8. The expected values shown are the arithmetic mean of the corresponding probability distributions which result from the Monte Carlo simulation. As such, they can be combined for the different areas to provide the expected levels of total production for different combined areas. (The required prices are relative to the landed costs of the crude oil and natural gas, implying that transportation charges to the market area are included in the costs.)

The relationship between higher expected price levels and incremental production does not represent the price sensitivity of oil and gas production for the overall areas. It indicates only the increased production that can be expected for a particular set of assumptions on size of resource base, field size distribution, and timing and size of the different lease sales—if the industry is convinced that the price levels of the landed crude and the landed gas at the points of reference will indeed materialize at the different levels shown after factors such as the regulatory climate and energy supply-and-demand conditions are taken into account.

In 1985 expected crude oil production from OCS areas is expected to range from 234 million barrels per year (or 0.64 million barrels per day) under a price scenario of $4.50 per barrel to 1038 million barrels per year (or 2.84 million barrels per day) under a price scenario of $18 per barrel. Expected gas production levels in 1985 will range from 1760 billion cubic feet per year (or 4.82 billion cubic feet per day) assuming a $.75 per thousand cubic feet price to 2963 billion cubic feet per year (or 8.12 billion cubic feet per day) assuming a $3 per thousand cubic feet price.

Supply Curves for New Offshore Areas

The information contained in Tables 4-7 and 4-8 on the expected production levels for crude oil and nonassociated gas in benchmark years at different price levels allows us to construct the supply curves showing how sensitive oil and

Table 4-7
Expected Production Levels for Crude Oil in Benchmark Years from Selected OCS Areas Leased or to Be Leased through 1978

Oil Production (10^6 bbl/yr)

Landed Price[a]

	$4.50/bbl			$7.50/bbl			$12/bbl			$18/bbl		
	1980	1985	1990	1980	1985	1990	1980	1985	1990	1980	1985	1990
N. Atlantic	19.13	19.30	18.91	25.14	42.44	27.50	25.92	43.80	28.38	25.93	43.94	28.48
Mid-Atlantic	33.21	56.77	36.78	44.30	76.77	49.82	44.96	77.99	50.61	45.01	78.12	50.70
S. Atlantic	2.72	6.69	4.38	8.95	21.72	14.29	9.53	23.24	15.29	9.55	23.35	15.37
Total	55.06	92.76	60.07	78.39	140.93	91.61	80.41	145.02	94.28	80.49	145.42	94.55
Gulf of Mexico												
E. Gulf MALFA	15.25	27.84	22.55	45.50	55.93	40.70	45.68	56.09	40.80	45.69	56.09	40.80
Central & West Gulf	19.10	70.53	76.76	44.98	140.69	128.87	45.08	141.22	129.32	45.09	141.24	129.33
Total	34.35	98.38	99.31	90.47	196.62	169.57	90.76	197.32	170.11	90.77	197.34	170.13
Pacific OCS												
S. California	1.00	2.60	3.71	56.0	154.70	134.00	59.40	165.37	141.90	59.60	166.40	142.62
Washington-Oregon	1.87	23.35	16.05	3.09	51.18	35.69	3.26	54.59	38.07	3.26	54.88	38.27
Total	2.87	25.95	19.76	59.09	205.88	169.69	62.66	219.96	179.97	62.86	221.28	180.89
Alaska OCS												
GOA East	—	—	—	—	56.34	40.76	—	75.17	54.89	—	76.30	55.76
GOA Kodiak	—	—	—	—	14.04	11.16	—	23.74	18.89	—	24.85	19.80
GOA S. Aleutian	—	—	—	—	—	—	—	0.37	1.35	—	0.71	2.25
Lower Cook Inlet	2.69	16.68	11.50	7.50	87.40	61.65	7.50	95.30	67.41	7.50	96.33	68.17
Bristol Basin	—	—	—	—	31.24	43.97	—	33.12	47.83	—	33.18	47.92
Bering Sea, Norton	—	—	—	—	18.84	23.79	—	28.51	34.54	—	29.02	35.08
Bering Sea, St. George	—	—	—	—	53.64	42.09	—	66.03	52.59	—	67.03	53.43
Chukchi Sea	—	—	—	—	1.31	1.14	—	1.31	5.60	—	1.31	5.86
Beaufort Sea	—	—	—	7.50	105.73	84.35	7.50	141.83	112.96	7.50	144.79	115.69
Total	2.69	16.68	11.50	7.50	368.54	308.90	7.50	465.39	396.06	7.50	473.52	403.96
Grand Total	94.97	233.77	190.64	235.45	911.97	739.77	241.33	1027.69	840.42	241.62	1037.56	849.53
10^6 bbl/day	0.26	0.64	0.52	0.65	2.50	2.03	0.66	2.82	2.30	0.66	2.84	2.33

[a]For Alaskan areas crude oil is assumed to be landed in California.

Table 4-8
Expected Production Levels for Nonassociated Gas in Benchmark Years from Selected OCS Areas Leased or to Be Leased Through 1978

Gas Production (10⁹ CF/yr)

Landed Price[a]

	$.75/10³ CF			$1.25/10³ CF			$2/10³ CF			$3/10³ CF		
	1980	1985	1990	1980	1985	1990	1980	1985	1990	1980	1985	1990
N. Atlantic	55.09	88.95	74.58	84.13	140.48	118.51	89.00	151.89	128.48	89.28	152.27	128.80
Mid-Atlantic	106.38	133.25	106.93	124.30	171.98	140.47	130.23	185.24	151.98	130.33	185.72	152.40
S. Atlantic	–	–	–	11.92	27.14	24.13	16.21	34.70	30.69	16.30	34.88	30.84
Total	161.47	222.20	181.51	220.35	339.60	283.10	253.44	371.83	311.16	235.91	372.87	312.04
Gulf of Mexico												
E. Gulf MAFLA	14.49	14.49	9.73	40.99	41.18	27.41	44.03	44.21	29.34	44.03	44.21	29.34
Central & West Gulf	939.96	1413.01	1140.48	1067.94	1650.72	1343.01	1078.77	1671.83	1361.15	1078.77	1671.83	1361.15
Total	954.44	1427.50	1150.21	1108.93	1691.90	1370.42	1122.80	1716.05	1390.49	1122.80	1716.05	1390.49
Pacific OCS												
S. California	66.74	88.56	69.23	90.25	131.99	106.17	95.76	143.38	116.11	96.72	146.04	118.46
Washington-Oregon	2.80	22.68	21.15	2.86	47.65	45.15	3.75	58.90	55.78	3.87	61.24	58.02
Total	69.60	111.24	90.38	93.11	179.65	151.32	99.50	202.28	171.89	100.58	207.28	176.48
Alaska OCS												
GOA East	–	–	–	–	–	–	–	73.56	71.97	–	112.57	110.98
GOA Kodiak	–	–	–	–	–	–	–	13.30	13.30	–	22.63	23.59
GOA S. Aleutian	–	–	–	–	–	–	–	–	–	–	0.42	3.33
Lower Cook Inlet	–	–	–	–	–	–	3.91	112.74	108.84	3.91	126.95	122.85
Bristol Basin	–	–	–	–	–	–	–	54.51	83.51	–	65.77	100.09
Bering Sea, Norton	–	–	–	–	–	–	–	–	–	–	19.38	23.38
Bering Sea, St. George	–	–	–	–	–	–	–	–	–	–	111.64	115.81
Chukchi	–	–	–	–	–	–	–	–	–	–	6.32	26.09
Beaufort Sea	–	–	–	–	–	–	–	–	–	–	200.94	201.05
Total	–	–	–	–	–	–	3.91	254.11	277.62	3.91	666.61	727.16
Grand Total	1185.51	1760.93	1422.09	1422.39	2211.14	1804.84	1461.65	2544.26	2151.16	1463.20	2962.80	2606.18
10⁹ CF/day	3.25	4.82	3.90	3.90	6.06	4.95	4.01	6.97	5.89	4.01	8.12	7.14

Source: Arthur D. Little, Inc., estimates.

[a]For Alaskan areas gas is assumed to be landed in California.

gas supplies resulting from the first round of lease sales are to expected future price levels.

These supply curves, derived by plotting the total production levels for each of the benchmark years on the horizontal axis against the corresponding prices on the vertical axis, are shown in Figure 4–16 *a* and *c*, respectively, for crude oil and nonassociated gas. Closer inspection shows all these supply curves to have significant "kinks." In the case of crude oil, the supply curve becomes almost vertical above a price of $7.50 per barrel, showing a high insensitivity of the expected supply to additional price increases. Also, the supply curve for the benchmark year 1990 lies to the left of the supply curve for the benchmark year 1985.

The occurrence of the kinks in the supply curves and the fact that the supply curve for 1990 turns out to lie to the left of the supply curve of 1985 can be explained by the fact that production levels realized in those benchmark years are only from fields found in areas assumed to be opened up through *the first round of lease sales* to be held through 1978. As discussed in the previous section, it is expected that the industry's efforts in those areas will concentrate on exploring the larger structures with the highest promise for large fields. Therefore, production from the areas leased first will be mainly from large, less costly fields, resulting in the relatively high price sensitivity over the lower part of the supply curves, as shown in Figure 4–16 *a* and *c*. The fact that field development will extend beyond 1980, but will be largely finished before 1985, explains why this price-sensitive part of the supply curve is so much steeper for 1980 than for 1985. The supply curve for 1990 lies to the left of the supply curve for 1985 because oil and gas fields found in the areas scheduled to be leased through 1978 will reach peak production around 1985 and be on the decline by 1990.

Assuming that lease sales will continue to be held through the 1980s, it can be expected that in the case of high price expectations the oil industry will continue to explore offshore areas to find more fields. Most of these fields will be more costly to develop than the fields found in the areas opened up through the first round of lease sales through 1978. The result will be, first, a supply curve for 1990 which will lie to the right of the supply curve of 1985 because the production decline of fields found subsequent to the first round of lease sales will be offset by the additional production from fields found subsequent to following rounds of lease sales. Second, the additional production from fields found subsequent to the rounds of lease sales following the first round will increase the production levels in 1985 and 1990, especially at higher levels of expected prices, smoothing out the kinks in the supply curves shown in Figure 4–16 *a* and *c*. The supply curve for 1980, based on the first round of lease sales through 1978, can be expected to change very little by subsequent rounds of lease sales, because of inherent delays between the time of a lease sale and the time of first production.

111

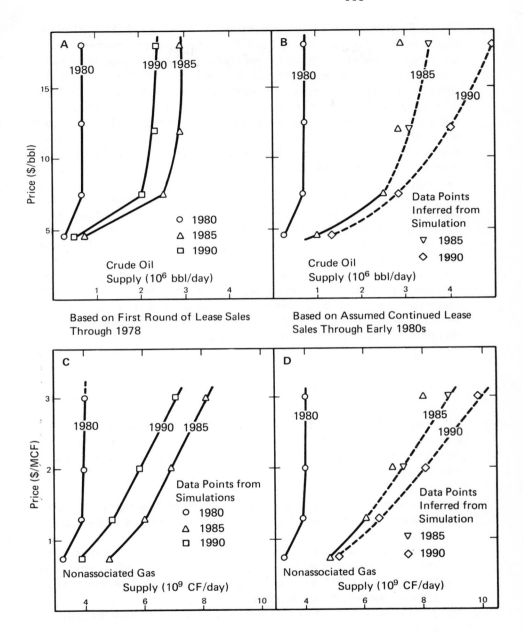

Figure 4–16. New Offshore Areas: Oil and Gas Supply Curves for Benchmark Years.

To illustrate the above observations, supply curves for 1985 and 1990 were drawn, as shown in Figure 4-16 *b* and *d*, for crude oil and nonassociated gas respectively, allowing for the additional production which one can expect to become available in 1985 and 1990, mostly from more costly fields. These supply curves are based on rough extrapolations from the supply curves shown in Figure 4-16 *a* and *c* and discussed above, and, as such, should only be considered as tentative. However, they do illustrate the shift to the right of the supply curve for 1990 due to additional production from fields found subsequent to later rounds of lease sales and the smoothing effect that this additional production is expected to have on the supply curves for 1985 and 1990.

Specifically, the supply curve for 1985 for oil, shown in Figure 4-16 *b*, still has a "kink" at a price level of around $7.50 per barrel. However, the supply curve for 1985 for gas is shown to have been completely "smoothed out" by the additional production. This difference between the gas supply curve and the oil supply curve for 1985 is explained by the fact that most of the undiscovered gas reserves are estimated by the USGS to be in the highly accessible Gulf of Mexico area, which will allow a more rapid field development and earlier peak production than in the case of oil, where more of the undiscovered reserves are estimated to be in more remote areas off the coasts of Alaska (see Table 3-4). For 1990, the oil and gas supply curves are shown to take on the smooth shape which they are supposed to have in traditional economic textbooks.

Production from Onshore and Existing Offshore Areas

To assess the potential impact of expected production from new OCS areas, a forecast was required, by state, of future potential production from onshore areas and existing offshore areas. These projections of production were made as follows:

1. Mean values for estimated undiscovered recoverable resources for 75 petroleum provinces as obtained from the USGS were assigned to the individual states.
2. Remaining revisions and extensions were calculated as follows: USGS Total Resources minus USGS Cumulative Production minus USGS Undiscovered Resources minus API Reserves (as of December 31, 1974) equals Revisions and Extensions to Proved Reserves.
3. A high and low projection of total production were made by projecting separately:
   a. Production from existing reserves derived by declining 1974 production levels for the individual states at 10 percent per annum
   b. Production from revisions and extensions to reserves existing in 1974, using the national availability profile to obtain the production profile for extensions and revisions realized in any given year

c. Production from newly discovered reserves assuming an optimistic and a pessimistic discovery scenario

d. Production from extensions and revisions to newly discovered reserves

An *optimistic production forecast* was obtained by assuming that economic incentives would result in an increase in discovery rates relative to 1974 levels (see Figure 4-17). Under that scenario half (50 percent) of the undiscovered resources were assumed to be discovered within the next 25 years, and all the undiscovered resources were assumed to be discovered in the next 50 years. The future production levels for Alaska onshore (Prudhoe Bay) were prespecified. Projections of production from existing offshore areas consisted of estimates of declining production from 1975 production levels and of estimates of production from extensions and revisions to those reserves.

Under this scenario, the annual discovery rate for oil would grow from 300 million barrels per year in 1974 to a peak of 960 million barrels per year in 1985, and decline thereafter (Figure 4-17). However, in spite of the increase in discovery rate, total production of crude oil and natural gas liquids from all areas under this scenario would continue to decline from the level of 9.65 million barrels per day in 1975 to 8.75 million barrels per day in 1985, followed by a period of steady growth in production capacity at a rate of about 2 percent per year to a level of about 9.8 million barrels per day in 1990 (Figure 4-18). The initial continuing decline in daily production between 1975 and 1985 is explained by the fact that at least 10 years will be required before increases in accumulated daily production capacity resulting from increased newly discovered reserves will overtake the decline in daily production capacity from reserves discovered prior to 1975. Also, production by extended oil recovery methods was assumed to make significant contributions to overall production between 1980 and 1985.

Under this optimistic production scenario, the daily production capacity for gas would continue to decline until 1980, in spite of increases in discovery rates from the 1975 level of 3.75 trillion cubic feet per year to a peak of 8.8 trillion cubic feet per year in 1985 (Figure 4-19). Following the year 1980, production capacity of nonassociated and associated gas would start to grow slowly to around 55 trillion cubic feet per day in 1985 and 1990 from 51 trillion cubic feet per day in 1980. The earlier turnaround in production capacity by increases in annual discoveries, then shown for oil, would reflect the expected response from industry if prices for natural gas were allowed to rise considerably relative to 1975 price levels.

A *pessimistic production forecast* was obtained by assuming that a lack of economic incentives would result in relatively low future annual discovery rates, remaining at approximately the same level as realized in 1974. The future production levels for Alaska onshore (Prudhoe Bay) were prespecified. Projections of production from existing offshore areas were obtained by decreasing 1975 production levels and by estimating increases in productive capacity through extensions and revisions to reserves in those areas.

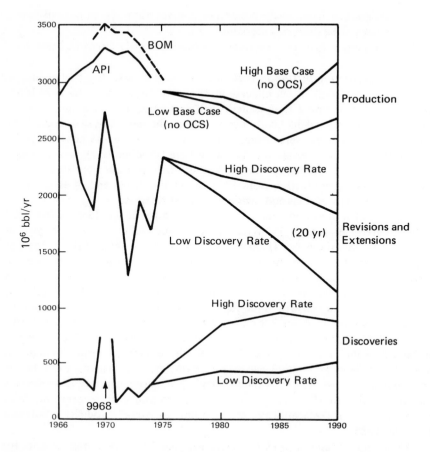

Figure 4-17. High and Low Base Case Projections of Discoveries, Revisions, and Extensions and Production from Onshore and Existing Offshore Areas of the United States—Oil.

Under this scenario, daily production of crude oil and natural gas liquids would decrease from a level of 9.6 million barrels per day in 1975 at an average rate of 3½ percent per year to 6.75 million barrels per day in 1985 (Figure 4-18). Starting around 1985, the production capacity would begin to increase again but only because of oil production from the reserves in the onshore areas of Alaska (Prudhoe Bay). Without this incremental production from Alaska, production would bottom out at 5½ million barrels a day in 1990. Extended oil recovery methods were assumed *not* to make any significant contribution to overall production.

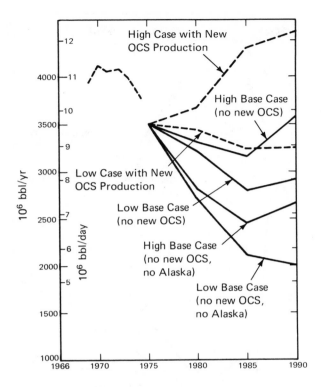

Source: Arthur D. Little, Inc., estimates.

Figure 4-18. Oil and Natural Gas Liquids—High and Low Projections of Production from Onshore and Existing Offshore Areas *with* and *without* Production from *New* OCS Areas of the United States.

Production of associated and nonassociated gas from existing and newly discovered fields onshore and from already discovered fields offshore would continue to decline from its level of 58 billion cubic feet per day in 1975 to 47 billion cubic feet per day in 1980, 44 trillion cubic feet per day in 1985, and 37 billion cubic feet per day in 1990, declining at an average rate of 3½ percent per year between 1975 and 1990 (see Figure 4-20).

The potential impact of future production from new OCS areas on the overall production capacity of the United States is shown by adding the projections for OCS production under high and low price scenarios to the low and high projections from onshore areas and from existing offshore areas discussed above. The price scenarios assumed for expected oil and gas production from

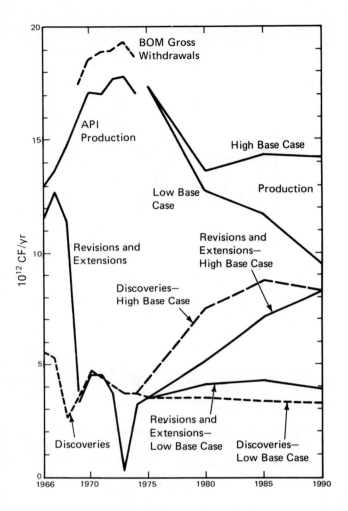

Source: Arthur D. Little, Inc., estimates.

Figure 4–19. Gas—High and Low Base Case Projections of Discoveries, Revisions and Extensions, and Production from Onshore and Existing Offshore Areas of the United States.

the new OCS areas were $12 per barrel and $1.25 per MCF, respectively, at the wellhead.

Total Future Potential Production of Crude Oil and Natural Gas Liquids. Combinations of the optimistic and pessimistic production forecasts for onshore

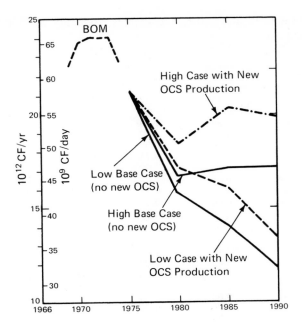

Source: Arthur D. Little, Inc., estimates.

Figure 4-20. Gas and Associated Gas—High and Low Projections of Production from Onshore and Existing Offshore Areas *With* and *Without* Production from *New* OCS Areas of the United States.

areas and existing offshore areas with the production forecasts for new OCS areas were made under high and low price scenarios, with the following results:

Under the optimistic/high price scenario, the following were found:

1. Total oil and natural gas liquids production would increase from a level of 9.6 million barrels per day in 1975 to about 10 million barrels per year in 1980, 11.6 million barrels per day in 1985, and 12.3 million barrels per day in 1990 (see Table 4-9).

2. Relative contribution to total domestic production from offshore areas would grow from about 17½ percent in 1975 to about 31 percent in 1985 (see Table 4-9).

3. About 36 percent of all OCS production in 1985 would come from areas offshore Alaska, about 24 percent from areas offshore the Pacific Coast, about 12 percent from areas offshore the Atlantic Coast, and 28 percent from areas in the Gulf of Mexico.

Table 4–9

Projections of Crude Oil and Natural Gas Liquids Production by Producing Region (10^6 bbl/day)

	Optimistic Case (1)				
	1974	1975	1980	1985	1990
Lower 48, old	8.67	7.92	5.89	4.24	3.38
Lower 48, new	0.00	0.00	0.71	1.98	3.52
Gulf of Mexico, old	1.36	1.23	0.88	0.33	0.30
Gulf of Mexico, new	0.00	0.00	0.36	0.71	0.60
Atlantic, new	0.00	0.00	0.24	0.43	0.29
Pacific, old	0.23	0.21	0.17	0.16	0.15
Pacific, new	0.00	0.00	0.27	0.49	0.50
Alaska onshore, new	0.03	0.08	1.37	1.92	2.47
Alaska offshore, old	0.16	0.15	0.09	0.05	0.03
Alaska offshore, new	0.00	0.00	0.02	1.30	1.11
Total	10.45	9.59	10.00	11.61	12.35

	Pessimistic Case (2)				
Lower 48, old	8.67	7.92	5.89	4.24	3.38
Lower 48, new	0.00	0.00	0.44	1.02	1.74
Gulf of Mexico, old	1.36	1.23	0.89	0.33	0.30
Gulf of Mexico, new	0.00	0.00	0.19	0.41	0.39
Atlantic, new	0.00	0.00	0.17	0.28	0.18
Pacific, old	0.23	0.21	0.18	0.16	0.15
Pacific, new	0.00	0.00	0.06	0.15	0.07
Alaska onshore, new	0.03	0.08	1.37	1.92	2.47
Alaska offshore, old	0.16	0.15	0.09	0.05	0.03
Alaska offshore, new	0.00	0.00	0.01	0.09	0.07
Total	10.45	9.59	9.28	8.65	8.78

(1) Assumptions:
1. For onshore areas other than Alaska, annual discoveries will increase at a rate of 11% per year from 300 million barrels of recoverable reserves in 1974 to 950 million barrels in 1985, and they will decline thereafter.
2. Production from onshore areas of Alaska will be as shown, mainly reflecting increases in production from the Prudhoe Bay area.
3. Production from offshore reserves producing in 1975 will continue to decline as shown.
4. For new OCS areas expected production will be as found with a $12/bbl wellhead price for oil and a $1.25/MCF wellhead price for gas assuming an accelerated lease sale schedule through 1978.
5. Extended oil recovery methods will start to contribute significantly to overall production between 1980 and 1985.

(2) Assumptions:
1. For onshore areas other than Alaska, annual discoveries will increase at a rate of only 3.5% per year from 300 million barrels of recoverable reserves in 1974 to 500 million barrels of recoverable reserves in 1990.
2. Production from onshore areas of Alaska will be as shown, reflecting mainly increases in production from the Prudhoe Bay area.
3. Production from offshore reserves, producing in 1975, will continue to decline as shown.
4. For new OCS areas expected production will be as found with a $4.50/bbl wellhead price for oil and a $0.75/MCF price for gas assuming an accelerated lease sale schedule through 1978.
5. Extended oil recovery methods will continue to contribute only marginally to overall production.

4. The contribution of total offshore production of crude oil and natural gas liquids would change between 1975 and 1985 (see Table 4-9):
   a. For Alaska, from 5½ to 36 percent, or from 0.15 million barrels per day to 1.35 million barrels per day
   b. For the Pacific, from 13 to 24 percent, or from 0.21 million barrels per day to 0.85 million barrels per day
   c. For the Atlantic, from 0 to 12 percent, or from 0.0 million barrels per day to 0.43 million barrels per day
   d. For the Gulf of Mexico, from 78 to 28 percent, or from 1.23 million barrels per day to 1.04 million barrels per day

Under the pessimistic/low price scenario, the following were found:

1. Total production of oil and natural gas liquids would decrease slightly from a level of 9.6 million barrels per day in 1975 to about 9.3 million barrels per day in 1980 and 8.7 million barrels per day in 1985, and then increase to 8.8 million barrels per day in 1990
2. Relative contribution to the total domestic production from offshore areas would grow from about 17½ percent in 1975 to about 19½ percent in 1985.
3. About 8 percent of all OCS production in 1985 would come from Alaska, 33 percent from the Pacific Coast areas, 16 percent from areas off the Atlantic Coast, and 43 percent from areas in the Gulf of Mexico.
4. The contribution to total offshore production of crude oil and natural gas liquids would change between 1975 and 1985 (see Table 4-9):
   a. For Alaska, from 5½ to 8 percent, or from 0.15 million barrels per day to 0.14 million barrels per day
   b. For the Pacific, from 13 to 33 percent, or from 0.21 million barrels per day to 0.57 million barrels per day
   c. For the Atlantic, from 0 to 16 percent, or from 0 million barrels per day to 0.28 million barrels per day
   d. For the Gulf of Mexico areas, from 78 to 43 percent, or from 1.23 million barrels per day to 0.74 million barrels per day

Total Future Potential Production of Associated and Nonassociated Natural Gas. Combinations of the optimistic and pessimistic production forecasts for onshore areas and existing offshore areas with the production forecasts for new OCS areas were made under high and low price scenarios, which provided the following results.
Under the optimistic/high price scenario:

1. Total associated and nonassociated gas production would decrease from a level of 58.2 billion cubic feet per day in 1975 to about 50.6 billion cubic feet in 1980, and then increase to 55.6 billion cubic feet per day in 1985 and 54.3 billion cubic feet per day in 1990 (see Table 4-10).

Table 4–10

Projections of Associated and Nonassociated Natural Gas Production by Producing Regions (10^9 CF/day)

	1974	1975	1980	1985	1990
		Optimistic Case (1)			
Lower 48, old	49.30	46.11	30.88	21.07	12.49
Lower 48, new	0.00	0.00	6.71	16.72	27.26
Gulf of Mexico, old	12.53	11.40	7.76	5.03	1.87
Gulf of Mexico, new	0.00	0.00	3.24	5.07	4.13
Atlantic, new	0.00	0.00	0.78	1.25	0.98
Pacific, old	0.14	0.13	0.13	0.12	0.12
Pacific, new	0.00	0.00	0.54	1.03	0.77
Alaska onshore, new	0.34	0.34	0.34	4.00	5.48
Alaska offshore, old	0.24	0.22	0.16	0.11	0.08
Alaska offshore, new	0.00	0.00	0.03	1.19	1.17
Total	62.55	58.20	50.57	55.59	54.35
		Pessimistic Case (2)			
Lower 48, old	49.30	46.11	30.88	21.07	12.49
Lower 48, new	0.00	0.00	4.15	8.45	12.47
Gulf of Mexico, old	12.53	11.40	7.76	5.03	1.87
Gulf of Mexico, new	0.00	0.00	2.69	4.27	3.37
Atlantic, new	0.00	0.00	0.56	0.81	0.63
Pacific, old	0.14	0.13	0.13	0.12	0.12
Pacific, new	0.00	0.00	0.39	0.62	0.45
Alaska onshore, new	0.34	0.34	0.34	4.00	5.48
Alaska offshore, old	0.24	0.00	0.16	0.11	0.08
Alaska offshore, new	0.00	0.00	0.00	0.00	0.00
Total	62.55	58.20	47.22	44.49	36.98

(1) Assumptions:

1. For onshore areas other than Alaska, annual discoveries will increase at a rate of 9% per year from 3.75 trillion cubic feet of recoverable reserves in 1974 to 3.75 trillion cubic feet in 1985, and they will decline thereafter.
2. Production from onshore areas of Alaska will be as shown, mainly reflecting increases in production from the Prudhoe Bay area.
3. Production from offshore reserves producing in 1975 will continue to decline as shown.
4. For new OCS areas expected production will be as found with a $12/bbl wellhead price for oil and a $1.25/MCF wellhead price for gas assuming an accelerated lease sale schedule through 1978.
5. Extended oil recovery methods will start to contribute significantly to overall production between 1980 and 1985.

(2) Assumptions:

1. For onshore areas other than Alaska, annual discoveries will decrease at a rate of 1% per year from 3.75 trillion cubic feet of recoverable reserves in 1974 to 3.2 trillion cubic feet in 1990.
2. Production from onshore areas of Alaska will be as shown, reflecting mainly increases in production from the Prudhoe Bay area.
3. Production from offshore reserves, producing in 1975, will continue to decline as shown.
4. For new OCS areas expected production will be as found possible with a $4.50/bbl wellhead price for oil and a $0.75/MCF wellhead price for gas assuming an accelerated lease sale schedule through 1978.
5. Extended oil recovery methods will continue to contribute only marginally to overall production.

Source: Arthur D. Little, Inc., estimates.

2. Relative contribution to total domestic production for offshore areas would grow from about 21 percent in 1975 to about 25 percent in 1985.
3. About 73 percent of all OCS production in 1985 would come from the Gulf of Mexico areas, about 9 percent from areas offshore the Atlantic Coast, about 8 percent from areas offshore the Pacific Coast, and about 10 percent from areas offshore Alaska.
4. The contribution to total offshore production of associated and nonassociated natural gas would change between 1975 and 1985 (see Table 4-10):
   a. For Alaska, from 2 to 10 percent, or from 0.22 billion cubic feet per day to 1.3 billion cubic feet per day
   b. For the Pacific, from 1 to 8 percent, or from 0.13 billion cubic feet per day to 1.15 billion cubic feet per day
   c. For the Atlantic, from 0 to 9 percent, or from 0 million cubic feet per day to 1.25 billion cubic feet per day
   d. For the Gulf of Mexico, from 97 to 73 percent, or from 11.4 billion cubic feet per day to 10.1 billion cubic feet per day

Under the pessimistic/low price scenario:

1. Total production of associated and nonassociated gas would decrease significantly from a level of 58.2 billion cubic feet per day to 47.2 billion cubic feet per day in 1980, 44.5 billion cubic feet per day in 1985, and 37.0 billion cubic feet per day in 1990.
2. Relative contribution to the total domestic production from offshore areas would grow from about 21 percent in 1975 to about 25 percent in 1985.
3. About 85 percent of all OCS production in 1985 would come from the Gulf of Mexico, 7 percent from the areas off the Atlantic Coast, 7 percent from areas off the Pacific Coast, and 1 percent from offshore Alaska.
4. The contribution to total offshore production of associated and nonassociated natural gas would change between 1975 and 1985 (see Table 4-10):
   a. For Alaska, from 2 to 1 percent, or from 0.22 billion cubic feet per day to 0.11 billion cubic feet per day
   b. For the Pacific, from 1 to 7 percent or from 0.13 billion cubic feet per day to 0.74 billion cubic feet per day
   c. For the Atlantic, from 0 to 7 percent, or from 0 billion cubic feet per day to 0.81 billion cubic feet per day
   d. For the Gulf of Mexico, from 97 to 85 percent, or from 11.4 billion cubic feet per day to 9.3 billion cubic feet per day

Capital Requirements

The capital expenditures required for exploration and development of each OCS area were determined as a part of the cost and production projections. These expected annual and cumulative capital expenditure projections for the various OCS areas are summarized in Table 4-11 for four different oil and gas price levels. The price levels are for oil and gas, respectively: (1) $4.50 per barrel, $0.75 per thousand cubic feet; (2) $7.50 per barrel, $1.25 per thousand cubic feet; (3) $12 per barrel, $2 per thousand cubic feet; and (4) $18 per barrel, $3 per thousand cubic feet. As the prices increase, smaller fields are developed, and the increases in capital expenditures with increasing prices reflect the additional field developments. The capital requirement projections explicitly exclude lease costs to the federal government. These lease costs historically have represented a significant portion of the total costs of developing a field.

The annual capital expenditures are expected to reach a maximum around 1980 followed by rapid decline as fields are completed to about 10 percent of the 1980 expenditure in 1985 and still less in 1990. Considering the price scenario of $12 per barrel for oil and $2 per thousand cubic feet for gas in Table 4-11, the total expected capital requirements for all OCS areas are $2.7 billion in 1980, $144 million in 1985, and $19 million in 1990 with a cumulated required investment from the present through 1990 of $13.8 billion (in 1975 dollars). The annual capital expenditures appear significant compared to the capacity of the oil and gas industry for capital generation for exploration and development. It can be estimated that the oil and gas industry invested about $4 billion in 1974 for exploration and development. Hence, it must be concluded that the development of the OCS will require a significant effort for the oil and gas industry during the peak years around 1980. If the prices increase, to $18 per barrel for oil and $3 per thousand cubic feet for gas, only small additional investments will be required in 1980 with an increase from $2.7 billion to $2.9 billion in the capital requirements. The breakdown of expenditures is described above in the section Total Costs for Exploration and Development for Individual Fields.

The cumulated capital requirements for the period through 1990 total $13.8 billion (under the $12 per barrel and $2 per thousand cubic feet price scenario). Compared to the total capital market, this amount appears very small and corresponds to less than 0.01 percent of the total GNP over the period.[2] When compared to other expected energy-related investments which have been estimated to be about $1 trillion through 1990, the OCS development also appears reasonably small.

The total capital investment which will be required for exploration and development of the OCS is very uncertain, and will vary extensively with the amount of oil and gas that will be located. If the amounts of oil and gas which are found are small, exploration will be pursued less vigorously and perhaps will be terminated early, and small development costs will be required. If the amounts found are large, the development activities which will be required are also large. Figure 4-21 presents the uncertainties for the six consolidated OCS areas. For each area, the capital which will be required for exploration and development

Table 4-11
Annual and Cumulative Capital Expenditures for Oil and Gas Production in OCS Areas ($ millions)

	$4.50/bbl, $0.75/10³ CF				$7.50/bbl, $1.25/10³ CF				$12/bbl, $2/10³ CF				$18/bbl, $3/10³ CF			
	1980	1985	1990	Cumulative Through 1990	1980	1985	1990	Cumulative Through 1990	1980	1985	1990	Cumulative Through 1990	1980	1985	1990	Cumulative Through 1990
N. Atlantic	46.65	0	0	272.46	92.10	0	0	480.58	116.13	0	0	576.48	118.60	0	0	584.94
Mid-Atlantic	82.68	0	0	474.93	136.02	0	0	755.45	161.61	0	0	852.08	164.06	0	0	860.92
S. Atlantic	10.49	0	0	42.93	52.47	0	0	228.29	68.28	0	0	304.46	69.64	0	0	309.88
Total	139.81	0	0	790.32	280.58	0	0	1464.32	346.01	0	0	1733.00	352.30	0	0	1755.73
Gulf of Mexico																
E. Gulf MAFLA	32.16	8.01	1.32	334.71	53.52	8.01	1.32	810.14	53.52	8.01	1.32	830.90	53.52	8.01	1.32	831.26
Central & West Gulf	337.41	51.52	17.91	2199.62	700.75	64.19	17.91	3814.51	741.84	65.44	17.91	3978.64	742.01	65.44	17.91	3979.89
Total	369.57	59.54	19.23	2534.33	754.27	72.20	19.23	4624.66	795.36	73.45	19.23	4809.54	795.53	73.45	19.23	4811.15
Pacific OCS																
S. California	133.03	0	0	882.36	230.30	0.11	0	1432.91	260.62	0.11	0	1567.96	270.17	0.11	0	1602.65
Washington-Oregon	55.54	0	0	161.66	145.91	0	0	413.47	169.83	0	0	500.90	174.72	0	0	517.81
Total	188.57	0	0	1044.02	376.21	0.11	0	1846.37	430.44	0.11	0	2068.86	444.90	0.11	0	2120.46
Alaska OCS																
GOA East	—	—	—	—	118.55	0.98	0	582.95	192.66	0.98	0	923.73	216.95	0.98	0	1070.67
GOA Kodiak	—	—	—	—	30.68	0.08	0	142.58	60.72	0.08	0	286.22	70.56	0.08	0	348.26
GOA S. Aleutian	—	—	—	—	—	—	—	—	1.37	2.63	0	19.12	5.33	4.89	0	51.24
Lower Cook Inlet	44.18	0.48	0	130.32	201.35	0.79	—	731.80	251.17	0.79	0	983.46	311.69	0.79	0	1157.63
Bristol Basin	—	—	—	—	34.85	21.96	0	333.64	55.88	26.94	0	466.53	66.70	28.92	0	584.37
Bering Sea, Norton	—	—	—	—	37.02	8.52	0	246.53	57.16	11.05	0	389.37	66.22	11.84	0	453.13
Bering Sea, St. George	—	—	—	—	116.26	1.74	0	479.87	139.93	1.74	0	626.86	181.59	1.74	0	840.87
Chukchi Sea	—	—	—	—	2.37	0	0	12.34	4.62	10.87	0	64.69	11.58	17.24	0	125.07
Beaufort Sea	44.18	0.48	0	130.32	276.19	14.99	0.23	1008.33	358.42	14.99	0.23	1401.83	419.50	16.10	0.23	1692.90
Total	44.18	0.48	0	130.32	817.26	49.06	0.23	3538.04	1121.93	70.07	0.23	5161.73	1350.13	82.58	0.23	6324.12
Grand Total	742.13	60.02	19.23	4498.98	2228.32	121.37	19.47	11,473.38	2693.75	143.63	19.47	13,773.18	2942.86	156.14	19.47	15,101.47

Source: Arthur D. Little, Inc., estimates.

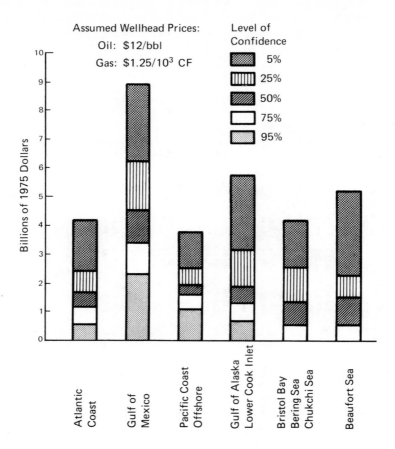

Source: Arthur D. Little, Inc., estimates.

Figure 4-21. Estimates of Required Total Capital Expenditures at Different Confidence Levels for Exploration Drilling and Field Development in Areas Leased or to Be Leased through 1978 on the Outer Continental Shelf of the United States.

may vary extensively for different levels of confidence. Hence, for the Atlantic Coast, the likelihood is 95 percent that the capital requirements will be at most $0.5 billion; 75 percent that it will be $1.1 billion or less; 50 percent that it will be $1.7 billion or less; 25 percent that it will be $2.4 billion; and 5 percent that it will be at most $4.2 billion. Between the confidence levels of 75 and

25 percent (which accounts for 50 percent of the expectations), the total capital expenditures for the Atlantic Coast increase from at most $1.1 billion to $2.4 billion, a range of 1 to 2.18. For the other OCS areas, the uncertainties vary similarly, as indicated in Figure 4-21.

Notes

1. United States Department of Interior, "Final Environmental Impact Statement: Proposed Increase in Oil and Gas Leasing on the Outer Continental Shelf."

2. "Capital Needs and Policy Choices in the Energy Industries," report submitted to the Federal Energy Administration, October 1974, by Arthur D. Little, Inc. (C-77389).

Appendix A
Resource Distributions for Oil and Gas by OCS Area

Table A-1

Resource Base Size Distribution, Gas

(10^12 CF of Recoverable Reserves)

				Cumulative Percentiles					
	0	*1*	*5*	*25*	*50*	*75*	*95*	*99*	*100*	*P(Dry)*
1. North Atlantic	0.0	2.4	3.0	5.0	6.8	9.0	15.0	20.0	30.0	0.40
2. Mid-Atlantic	0.0	2.4	3.0	5.0	6.5	9.0	15.0	21.0	30.0	0.30
3. South Atlantic	0.0	0.6	0.8	1.2	1.6	2.0	3.0	4.0	5.4	0.60
4. Eastern Gulf	0.0	0.35	0.5	0.85	1.25	1.8	3.0	4.4	6.5	0.30
5. Central Western Gulf	0.0	12.5	17.0	28.0	38.0	55.0	92.0	125.0	175.0	0.0
6. So. California } 6A. Santa Barbara	0.0	1.075	1.438	2.043	2.573	3.330	4.661	5.838	6.5	0.0
7. Washington, Oregon, & No. California	0.0	0.0	0.0	0.201	0.430	0.883	2.412	3.421	5.0	0.40
8. Gulf of Alaska, East	0.0	0.1	0.6	1.5	3.0	6.0	15.0	29.0	5.0	0.30
9. Gulf of Alaska, Kodiak	0.0	0.06	0.12	0.45	1.0	2.0	6.0	13.5	30.0	0.60
10. Gulf of Alaska, Aleutian Shelf	0.0	0.07	0.16	0.20	0.30	0.5	1.0	1.6	2.5	0.80

Table A–1 – *Continued*

Resource Base Size Distribution, Gas

(10^{12} CF of Recoverable Reserves)

11.	Lower Cook Inlet	0.0	0.8	1.0	1.6	2.2	3.0	4.5	6.3	8.4	0.0
12.	Outer Bristol Basin	0.0	0.2	1.6	2.25	3.0	4.0	6.0	8.0	11.0	0.50
13.	Bering Sea, Norton Basin	0.0	0.4	0.5	0.9	1.2	1.8	3.0	4.3	6.5	0.40
14.	Bering Sea, St. George	0.0	1.5	2.0	3.8	5.5	8.2	15.0	23.0	35.0	0.50
15.	Chukchi Sea (Hope Basin)	0.0	0.4	0.5	0.9	1.2	1.8	3.0	4.3	6.5	0.40
16.	Beaufort Sea	0.0	4.0	5.0	7.5	10.0	13.5	20.0	20.5	35.0	0.25

Table A-2
Resource Base Size Distribution, Oil
(10⁹ bbl of Recoverable Reserves)

					Cumulative Percentiles					
	0	1	5	25	50	75	95	99	100	P(Dry)
1. North Atlantic	0.0	0.4	0.6	1.0	1.35	1.8	3.0	4.0	6.0	0.40
2. Mid-Atlantic	0.0	0.8	1.0	1.67	2.2	3.0	5.0	7.0	10.0	0.30
3. South Atlantic	0.0	0.3	0.4	0.6	0.8	1.0	1.5	2.0	2.7	0.60
4. Eastern Gulf	0.0	0.35	0.5	0.85	1.25	1.8	3.0	4.4	6.5	0.30
5. Central Western Gulf	0.0	1.8	2.0	2.8	3.6	4.6	6.4	8.3	13.0	0.0
6. So. California										
6A. Santa Barbara	0.0	1.047	1.357	1.931	2.464	3.114	4.384	5.465	6.0	0.0
7. Washington, Oregon, & No. California	0.0	0.0	0.	0.223	0.451	0.751	1.311	1.795	2.0	0.40
8. Gulf of Alaska, East	0.0	0.1	0.2	0.5	1.0	2.0	5.0	10.0	20.0	0.30
9. Gulf of Alaska, Kodiak	0.0	0.02	0.04	0.15	0.33	0.67	2.0	4.5	10.0	0.60
10. Gulf of Alaska, Aleutian Shelf	0.0	0.03	0.05	0.10	0.16	0.25	0.5	0.8	1.3	0.90

Table A–2 – *Continued*

Resource Base Size Distribution, Oil

(*10⁹ bbl of Recoverable Reserves*)

11.	Lower Cook Inlet	0.0	0.4	0.5	0.8	1.1	1.5	2.25	3.15	4.2	0.0
12.	Outer Bristol Basin	0.0	0.2	0.5	0.8	1.25	1.75	3.0	4.2	7.0	0.50
13.	Bering Sea, Norton Basin	0.0	0.4	0.5	0.8	1.2	1.75	2.8	4.0	6.0	0.60
14.	Bering Sea, St. George	0.0	0.6	0.8	1.5	2.25	3.3	6.0	9.0	15.0	0.50
15.	Chukchi Sea (Hope Basin)	0.0	0.15	0.20	0.3	0.4	0.53	0.8	1.05	1.5	0.70
16.	Beaufort Sea	0.0	1.6	2.0	3.0	4.0	5.5	8.0	10.6	14.0	0.25

**Appendix B
Minimum Required Prices and Total Investment
Costs by Field Size for Each OCS Area**

Table B-1
Well Productivity, Minimum Required Price as a Function of Field Size, Oil
(Required Rate of Return: 15%) ($/bbl)[a]

| Area | Well Productivity (bbl/day) | Field Size (10⁶ bbl) | | | | | | | | | Water Depth (ft) | Distance to Shore (mi) | No. of Years Delay[b] |
		5	15	45	90	150	350	750	1400	2000			
Atlantic	500	43.34	18.34	10.08	7.81	7.50	6.59	5.71	5.65	5.78	400	75	4
	2500	41.02	17.17	8.85	6.74	5.82	4.60	4.06	3.78	3.57	400	75	4
	10,000	40.63	16.94	8.70	6.49	5.54	4.34	3.56	3.15	3.10	400	75	4
Gulf of Mexico	500	32.53	13.92	7.78	6.10	5.81	5.09	4.68	4.80	4.98	400	75	3
	2500	30.76	13.02	6.87	5.31	4.63	3.71	3.26	3.02	2.90	400	75	3
	10,000	30.43	12.81	6.72	5.09	4.40	3.49	2.88	2.55	2.50	400	75	3
California	500	45.77	19.44	10.50	7.47	6.91	5.91	5.26	5.21	5.32	600	15	4
	2500	43.37	18.34	9.31	6.48	5.27	4.12	3.82	3.63	3.43	600	15	4
	10,000	43.02	18.12	9.16	6.24	5.01	3.89	3.35	3.02	3.01	600	15	4
Washington-Oregon	500	46.09	19.59	10.58	7.51	6.94	5.93	5.27	5.22	5.33	600	15	4
	2500	43.67	18.48	9.39	6.52	5.29	4.13	3.83	3.63	3.44	600	15	4
	10,000	43.32	18.26	9.24	6.28	5.04	3.90	3.36	3.03	3.02	600	15	4
Gulf of Alaska	500	114.96	45.86	22.99	15.59	14.03	11.61	10.10	9.77	9.90	400	25	5
	2500	108.95	42.98	19.63	12.74	9.84	7.17	6.44	6.03	5.67	400	25	5
	10,000	108.36	42.68	19.44	12.25	9.26	6.61	5.44	4.80	4.79	400	25	5

	500	78.14	32.43	17.41	12.05	10.51	8.70	7.58	7.44	7.57	200	15	5
Lower Cook Inlet	2500	74.16	30.18	14.23	9.30	7.19	5.22	4.60	4.28	4.04	200	15	5
	10,000	73.66	29.91	14.06	8.81	6.61	4.66	3.77	3.33	3.29	200	15	5
	500	109.88	42.19	20.27	14.32	12.67	10.56	9.05	8.88	9.04	200	75	5
Bering Sea	2500	103.94	39.06	16.70	11.21	8.86	6.36	5.38	4.92	4.64	200	75	5
	10,000	103.26	38.71	16.51	10.66	8.20	5.73	4.42	3.82	3.75	200	75	5
	500	156.55	60.04	28.71	19.32	16.66	13.49	11.63	11.33	11.50	300	15	5
Beaufort Sea	2500	148.14	55.79	23.78	15.09	11.42	8.05	6.98	6.46	6.08	300	15	5
	10,000	147.24	55.30	23.47	14.31	10.51	7.17	5.70	4.98	4.92	300	15	5

[a] 1975 dollars.

[b] Number of years delay after date of lease acquisition until first production is generated.

Table B-2
Well Productivity, Minimum Required Rate of Return as a Function of Field Size, Gas
(Required Rate of Return: 15%) ($/MCF)[a]

Area	Well Productivity (10⁶ CF/day)	Field Size (10³ CF)									Case Assumptions		
		50	100	250	500	1000	2500	5000	10000	20000	Water Depth (ft)	Distance to Shore (mi)	No. of Years Delay[b]
Atlantic	20	4.45	2.54	1.37	0.97	0.77	0.66	0.60	0.54	0.55	400	75	4
	50	4.43	2.53	1.36	0.96	0.75	0.61	0.51	0.48	0.45	400	75	4
	100	4.42	2.52	1.36	0.95	0.74	0.60	0.51	0.44	0.42	400	75	4
Gulf of Mexico	20	3.36	1.93	1.06	0.76	0.61	0.53	0.48	0.45	0.48	400	75	3
	50	3.35	1.92	1.05	0.75	0.59	0.49	0.42	0.39	0.38	400	75	3
	100	3.34	1.91	1.05	0.75	0.59	0.49	0.42	0.36	0.35	400	75	3
California	20	4.82	2.80	1.51	0.97	0.69	0.54	0.51	0.46	0.46	600	15	4
	50	4.80	2.79	1.50	0.96	0.67	0.50	0.42	0.40	0.39	600	15	4
	100	4.79	2.78	1.50	0.95	0.67	0.49	0.42	0.36	0.36	600	15	4
Washington-Oregon	20	4.88	2.85	1.55	0.99	0.70	0.55	0.51	0.46	0.46	600	15	4
	50	4.87	2.84	1.54	0.97	0.68	0.50	0.42	0.41	0.39	600	15	4
	100	4.86	2.83	1.53	0.97	0.68	0.49	0.42	0.36	0.36	600	15	4
Gulf of Alaska	20	12.16	6.83	3.46	2.06	1.35	0.97	0.88	0.78	0.76	400	25	5
	50	12.14	6.82	3.45	2.03	1.30	0.86	0.69	0.64	0.60	400	25	5
	100	12.13	6.81	3.45	2.03	1.30	0.84	0.67	0.56	0.54	400	25	5

Lower Cook Inlet	20	8.15	4.64	2.38	1.40	0.92	0.64	0.57	0.50	0.49	200	15	5
	50	8.14	4.62	2.37	1.38	0.87	0.56	0.44	0.41	0.38	200	15	5
	100	8.13	4.61	2.36	1.37	0.86	0.54	0.43	0.35	0.34	200	15	5
Bering Sea	20	11.14	5.98	2.84	1.77	1.24	0.94	0.81	0.71	0.71	200	75	5
	50	11.13	5.97	2.82	1.74	1.18	0.84	0.66	0.59	0.55	200	75	5
	100	11.11	5.96	2.82	1.74	1.18	0.82	0.64	0.52	0.49	200	75	5
Beaufort Sea	20	16.44	8.93	4.26	2.47	1.58	1.07	0.94	0.83	0.81	300	15	5
	50	16.40	8.90	4.23	2.43	1.50	0.93	0.72	0.65	0.60	300	15	5
	100	16.37	8.88	4.22	2.41	1.48	0.90	0.69	0.56	0.53	300	15	5

[a]1975 dollars.

[b]Number of years delay after discovery date of well until first production is generated.

Table B-3
Investment as a Function of Field Size, Oil
(Average Well Productivity of Oil: 2500 bbl/day)[a]
(millions of 1975 dollars)

Area/Investment Type	Field Size (in 10^6 bbl of Recoverable Reserves)							
	15	*45*	*90*	*150*	*350*	*750*	*1400*	*2000*
Atlantic								
Exploration Wells (4)	8.4	8.4	8.4	8.4	8.4	8.4	8.4	8.4
Platform Construction and Installation	13.5	17.7	23.9	32.2	59.9	126.3	206.3	342.5
Development Wells	0.6	0.6	3.7	7.3	19.5	44.5	84.8	121.5
Platform Equipment	3.2	8.6	16.1	25.7	55.7	118.3	195.3	323.9
Pipeline to Shore	16.7	37.3	38.4	39.7	69.2	88.2	128.4	158.2
Gathering Lines	0.8	0.8	0.8	0.8	0.9	1.3	2.0	3.5
Onshore Terminal	2.1	6.3	12.9	21.3	49.2	103.8	188.4	262.8
Total Development	36.9	71.3	95.8	127.0	254.4	482.4	805.2	1212.4
Annual Production Cost	1.9	4.0	6.5	9.3	16.9	33.7	62.9	89.8
Gulf of Mexico[b]								
Exploration Wells (4)	8.0	8.0	8.0	8.0	8.0	8.0	8.0	8.0
Platform Construction and Installation	9.8	12.6	16.8	30.5	66.5	147.2	268.8	385.6
Development Wells	1.6	8.6	19.1	32.4	78.4	169.9	318.4	456.0
Platform Equipment	2.8	7.7	14.8	24.3	57.9	125.9	235.6	337.4
Pipeline to Shore	14.0	31.1	32.4	33.8	61.2	74.8	94.8	108.0
Gathering Lines	0.5	0.5	0.5	0.9	1.6	4.1	6.4	9.4
Onshore Terminal	2.2	6.6	13.4	22.2	51.6	108.6	188.4	239.6
Total Development	30.9	67.1	97.0	144.1	317.2	630.5	1112.4	1536.0
Annual Production Cost	1.4	3.2	5.4	9.1	21.9	46.4	82.2	106.0
California								
Exploration Wells (4)	8.0	8.0	8.0	8.0	8.0	8.0	8.0	8.0
Platform Construction and Installation	16.5	21.6	29.3	39.4	73.2	154.7	252.7	419.5
Development Wells	0.6	0.6	3.0	6.0	16.1	36.5	69.4	99.4
Platform Equipment	3.1	8.4	15.9	25.4	55.1	117.0	193.1	320.3
Pipeline to Shore	7.4	16.6	16.8	17.2	18.8	21.6	26.2	30.6
Gathering Lines	0.5	0.5	0.5	0.5	0.6	0.9	1.4	2.4
Onshore Terminal	2.1	6.3	12.9	21.3	49.2	103.8	188.4	262.8
Total Development	30.2	54.0	78.4	109.8	213.0	434.5	731.2	1135.0
Annual Production Cost	1.5	3.5	5.6	8.1	15.1	30.7	57.4	82.0

Table B–3 *(Continued)*

Area/Investment Type	Field Size (in 10^6 bbl of Recoverable Reserves)							
	15	45	90	150	350	750	1400	2000
Washington-Oregon								
Exploration Wells (4)	8.0	8.0	8.0	8.0	8.0	8.0	8.0	8.0
Platform Construction and Installation	16.5	21.6	29.3	39.4	73.2	154.7	252.7	419.5
Development Wells	0.6	0.6	3.0	6.0	16.1	36.5	69.4	99.4
Platform Equipment	3.1	8.4	15.9	25.4	55.1	117.0	193.1	320.3
Pipeline to Shore	7.8	17.2	17.6	18.0	19.6	22.8	28.2	33.0
Gathering Lines	0.7	0.7	0.7	0.8	0.8	1.2	1.8	3.0
Onshore Terminal	2.1	6.3	12.9	21.3	49.2	103.8	188.4	262.8
Total Development	30.8	54.8	79.4	110.9	214.0	436.0	733.6	1135.0
Annual Production Cost	1.5	3.5	5.7	8.1	15.1	30.7	57.4	82.1
Gulf of Alaska								
Exploration Wells (4)	21.2	21.2	21.2	21.2	21.2	21.2	21.2	21.2
Platform Construction and Installation	32.8	42.7	57.7	77.7	144.5	305.1	498.6	827.6
Development Wells	1.6	1.6	9.6	19.2	51.2	116.8	222.4	294.4
Platform Equipment	3.2	8.7	16.2	25.9	56.3	119.7	197.4	327.5
Pipeline to Shore	13.8	31.2	32.1	33.3	37.2	45.3	58.5	70.8
Gathering Lines	3.4	3.4	3.4	3.5	3.5	4.4	6.7	11.8
Onshore Terminal	2.4	7.4	14.8	24.6	57.0	120.2	217.6	302.2
Total Development	57.2	95.0	133.8	184.2	349.7	711.5	1201.2	1834.3
Annual Production Cost	3.1	6.5	10.0	13.9	23.9	47.8	89.6	128.1
Lower Cook Inlet								
Exploration Wells (4)	17.6	17.6	17.6	17.6	17.6	17.6	17.6	17.6
Platform Construction and Installation	11.3	14.8	19.8	26.7	49.6	104.7	171.3	284.2
Development Wells	1.6	1.6	9.6	19.2	51.2	116.8	222.4	318.4
Platform Equipment	3.2	8.7	16.2	25.9	56.3	119.7	197.4	327.5
Pipeline to Shore	11.1	24.6	25.2	25.8	28.2	33.0	41.1	48.3
Gathering Lines	3.4	3.4	3.4	3.4	3.5	4.3	6.6	11.6
Onshore Terminal	2.4	7.2	14.1	23.4	54.6	115.5	210.6	294.3
Total Development	33.0	60.3	88.3	124.4	243.4	494.0	849.4	1284.3
Annual Production Cost	2.7	5.7	8.7	12.0	20.8	41.7	78.1	111.7
Bering Strait								
Exploration Wells (4)	28.8	28.8	28.8	28.8	28.8	28.8	28.8	28.8
Platform Construction and Installation	11.3	14.8	19.8	26.7	49.6	104.7	171.3	284.2
Development Wells	1.9	1.9	11.1	22.2	59.3	135.1	257.3	368.3
Platform Equipment	3.2	9.0	16.8	26.9	58.6	124.4	205.3	340.5

Table B–3 *(Continued)*

Area/Investment Type	Field Size (in 10^6 bbl of Recoverable Reserves)							
	15	45	90	150	350	750	1400	2000
Pipeline to Shore	17.9	39.8	41.0	42.3	73.8	102.3	148.5	191.4
Gathering Lines	3.8	3.8	3.8	3.8	3.9	4.8	7.3	12.8
Onshore Terminal	2.7	8.4	16.8	27.9	64.8	137.4	251.4	350.4
Total Development	40.8	77.7	109.3	149.8	310.0	608.7	1041.1	1547.6
Annual Production Cost	4.0	7.4	11.0	15.0	25.0	49.2	92.2	131.9
Beaufort Sea								
Exploration Wells (4)	40.0	40.0	40.0	40.0	40.0	40.0	40.0	40.0
Platform Construction and Installation	23.2	30.2	40.9	54.9	102.1	215.5	352.1	584.6
Development Wells	2.4	2.4	14.4	28.8	76.8	152.7	333.6	477.6
Platform Equipment	3.5	9.3	17.7	28.2	61.8	131.1	216.6	359.3
Pipeline to Shore	14.4	32.2	32.6	33.2	35.4	39.8	46.8	53.4
Gathering Lines	4.4	4.4	4.4	4.5	4.5	5.5	8.4	14.9
Onshore Terminal	2.8	8.6	17.0	28.2	65.6	138.2	252.7	353.1
Total Development	50.7	87.1	127.0	177.8	346.2	682.8	1210.2	1842.9
Annual Production Cost	3.9	8.5	13.5	19.1	34.0	68.9	129.2	184.8

[a]Except Gulf of Mexico.

[b]Average well productivity is 50 bbl/day.

Table B–4
Investment as a Function of Field Size, Gas
(Average Well Productivity for Gas: 50 10^6 CF/day)[a]
(millions of 1975 dollars)

Area/Investment Type	Field Size (in 10^9 CF of Recoverable Reserves)							
	100	250	500	1000	2500	5000	10,000	20,000
Atlantic								
Exploration Wells (4)	8.4	8.4	8.4	8.4	8.4	8.4	8.4	8.4
Platform Construction and Installation	12.6	14.3	17.1	22.9	39.9	68.9	137.3	285.6
Development Wells	0.6	0.6	0.6	0.6	4.3	10.4	23.1	48.8
Platform Equipment	1.8	4.2	7.8	14.8	34.3	55.7	122.0	256.1
Pipeline to Shore	20.7	34.5	42.0	54.6	86.7	138.3	228.0	387.0
Gathering Lines	0.8	0.8	0.9	0.9	1.1	1.8	3.0	5.3
Onshore Terminal	0.0	0.0	0.0	0.0	0.0	0.0	0.0	0.0
Total Development	36.5	54.4	68.4	93.8	166.3	275.1	513.4	982.8
Annual Production Cost	1.8	3.9	7.:	13.1	29.9	56.3	113.5	227.8
Gulf of Mexico[b]								
Exploration Wells (4)	8.0	8.0	8.0	8.0	8.0	8.0	8.0	8.0
Platform Construction and Installation	9.3	10.6	12.6	16.8	34.5	72.3	147.7	298.8
Development Wells	0.6	0.6	1.0	3.6	11.6	24.6	50.5	102.5
Platform Equipment	1.8	4.2	7.8	14.7	30.9	66.9	138.9	282.5
Pipeline to Shore	19.2	31.2	38.3	49.8	79.0	123.2	201.4	357.6
Gathering Lines	0.5	0.5	0.6	0.6	1.0	1.9	3.7	7.4
Onshore Terminal	0.0	0.0	0.0	0.0	0.0	0.0	0.0	0.0
Total Development	31.4	47.1	60.3	85.5	157.0	288.9	542.2	1048.8
Annual Production Cost	1.5	3.4	6.3	11.9	29.0	58.3	116.6	233.1
California								
Exploration Wells (4)	8.0	8.0	8.0	8.0	8.0	8.0	8.0	8.0
Platform Construction and Installation	15.4	17.5	21.0	28.0	48.9	84.4	168.2	349.7
Development Wells	0.6	0.6	0.6	0.6	3.6	8.6	19.1	40.1
Platform Equipment	2.8	4.2	7.8	14.7	33.9	55.1	120.8	253.3
Pipeline to Shore	8.0	16.2	17.0	18.6	23.2	31.0	46.8	78.0
Gathering Lines	0.5	0.5	0.5	0.6	0.7	1.2	2.0	3.6
Onshore Terminal	0.0	0.0	0.0	0.0	0.0	0.0	0.0	0.0
Total Development	27.3	39.0	46.9	62.5	110.3	180.3	356.9	724.7
Annual Production Cost	1.6	3.6	6.5	12.0	27.2	51.9	104.7	210.2

Table B–4 *(Continued)*

Area/Investment Type	Field Size (in 10⁹ CF of Recoverable Reserves)							
	100	250	500	1000	2500	5000	10,000	20,000
Washington-Oregon								
Exploration Wells (4)	8.0	8.0	8.0	8.0	8.0	8.0	8.0	8.0
Platform Construction and Installation	15.4	17.5	21.0	28.0	48.9	84.4	168.2	349.7
Development Wells	0.6	0.6	0.6	0.6	3.6	8.6	19.1	40.1
Platform Equipment	3.8	4.2	7.8	14.7	33.9	55.1	120.8	253.3
Pipeline to Shore	8.8	18.0	18.8	20.4	25.0	32.8	48.6	79.8
Gathering Lines	0.7	0.8	0.8	0.8	0.9	1.6	2.6	4.5
Onshore Terminal	0.0	0.0	0.0	0.0	0.0	0.0	0.0	0.0
Total Development	29.3	41.1	49.0	64.5	112.3	182.5	359.3	727.4
Annual Production Cost	1.6	3.6	6.5	12.0	27.2	51.9	104.7	210.2
Gulf of Alaska								
Exploration Wells (4)	21.2	21.2	21.2	21.2	21.2	21.2	21.2	21.2
Platform Construction and Installation	30.5	34.5	41.5	55.1	96.5	116.6	331.9	689.9
Development Wells	1.6	1.6	1.6	1.6	11.2	27.2	60.8	126.4
Platform Equipment	3.6	4.3	7.9	15.1	34.8	43.8	123.3	258.7
Pipeline to Shore	18.3	36.9	38.4	41.1	49.8	64.2	93.3	150.9
Gathering Lines	3.5	3.6	3.6	3.7	3.8	3.3	9.5	14.8
Onshore Terminal	0.0	0.0	0.0	0.0	0.0	0.0	0.0	0.0
Total Development	57.5	80.9	93.0	116.6	196.1	255.1	618.8	1240.7
Annual Production Cost	2.7	5.8	9.9	17.0	35.8	65.8	133.5	268.7
Lower Cook Inlet								
Exploration Wells (4)	17.6	17.6	17.6	17.6	17.6	17.6	17.6	17.6
Platform Construction and Installation	10.4	11.9	14.2	18.9	33.1	40.0	113.9	236.9
Development Wells	1.6	1.6	1.6	1.6	11.2	27.2	60.8	128.0
Platform Equipment	3.6	4.3	7.9	15.1	34.8	43.8	123.3	258.7
Pipeline to Shore	13.8	27.9	28.8	30.6	35.7	44.4	61.8	96.3
Gathering Lines	3.4	3.4	3.4	3.5	3.6	3.1	8.9	13.6
Onshore Terminal	0.0	0.0	0.0	0.0	0.0	0.0	0.0	0.0
Total Development	32.8	49.1	55.9	69.7	118.4	158.5	368.7	733.5
Annual Production Cost	2.4	5.2	8.9	15.3	32.2	59.2	120.2	241.8
Bering Strait								
Exploration Wells (4)	28.8	28.8	28.8	28.8	28.8	28.8	28.8	28.8
Platform Construction and Installation	10.4	11.9	14.2	18.9	33.1	40.0	113.9	236.9
Development Wells	1.9	1.9	1.9	1.9	13.0	31.5	70.4	148.1
Platform Equipment	3.6	4.4	8.1	15.4	36.0	45.5	128.1	268.7

Table B–4 *(Continued)*

Area/Investment Type	Field Size (in 10⁹ CF of Recoverable Reserves)							
	100	250	500	1000	2500	5000	10,000	20,000
Pipeline to Shore	25.5	42.4	49.4	64.1	100.8	153.8	246.8	432.8
Gathering Lines	3.9	4.0	4.0	4.0	4.2	3.6	10.2	15.5
Onshore Terminal	0.0	0.0	0.0	0.0	0.0	0.0	0.0	0.0
Total Development	45.3	64.6	77.6	104.3	187.1	274.4	569.4	1102.0
Annual Production Cost	3.0	6.2	10.6	18.4	39.0	70.9	143.9	289.8
Beaufort Sea								
Exploration Wells (4)	40.0	40.0	40.0	40.0	40.0	40.0	40.0	40.0
Platform Construction and Installation	21.5	24.4	29.3	38.9	68.1	82.4	234.3	487.2
Development Wells	2.4	2.4	2.4	2.4	16.8	40.8	91.2	192.0
Platform Equipment	3.8	4.6	8.7	16.2	37.7	47.9	134.9	282.8
Pipeline to Shore	16.0	32.4	33.6	35.6	44.4	52.0	72.4	113.4
Gathering Lines	4.5	4.6	4.6	4.7	4.8	4.1	11.8	17.9
Onshore Terminal	0.0	0.0	0.0	0.0	0.0	0.0	0.0	0.0
Total Development	48.2	68.4	78.6	97.8	171.8	227.2	544.6	1093.3
Annual Production Cost	3.7	7.6	12.6	21.2	42.5	76.4	156.2	315.7

[a]Except Gulf of Mexico.

[b]Average well productivity is 20 MMCF/D.

Index

Index

About the Authors

Frederik W. Mansvelt Beck, as a member of the Energy Economics Section of Arthur D. Little, Inc., concentrates on computer modeling and economic analysis within the context of the exploration, development and production activities of the petroleum industry. He has been with ADL since 1973. His experience, prior to entering the consulting field, includes three years with Royal Dutch Shell working as a petroleum engineer on various assignments in the U.S.A., Europe, West Africa and Turkey. Mr. Mansvelt Beck received the M.S. in physics from the Technical University at Delft in the Netherlands and the M.S. in industrial administration from Carnegie Mellon University.

Karl M. Wiig is head of the Systems & Policy Analysis Unit within the Operations Research Section of Arthur D. Little, Inc. His work has centered around practical applications of mathematical modeling and systems science and on-line computer-based systems. His primary activities have been with improvement of operation of complex systems and analyses of long-range decision situations with development of optimal strategies. Mr. Wiig received the B.S. in mechanical engineering and the M.S. in instrumentation, both from Case Institute of Technology. He is the author and co-author of many papers on applied systems analysis, decision analysis, and control theory. He is a member of the Society for Industrial and Applied Mathematics, the Institute of Electrical and Electronic Engineers, the Institute of Management Sciences and the American Association for the Advancement of Science.